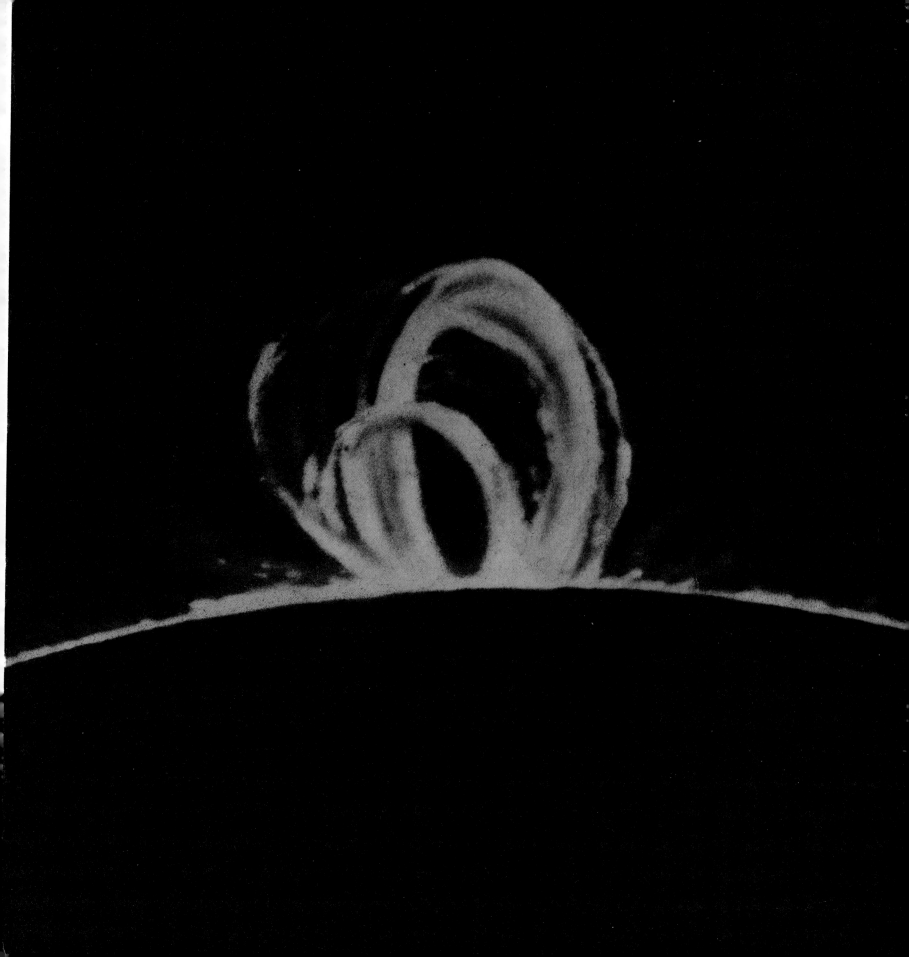

The
Smithsonian
Book
of the Sun

Fire of Life

Smithsonian Exposition Books

Distributed
by W.W. Norton & Company
New York, London

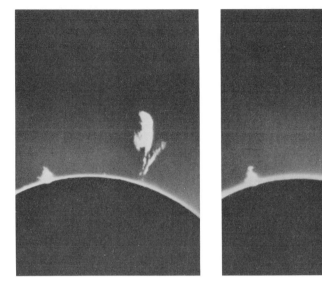

The Smithsonian Institution
Secretary: S. Dillon Ripley
Assistant Secretary, Public Service:
Julian T. Euell

Smithsonian Exposition Books
Director: Glen B. Ruh
Editors: Joe Goodwin, Alexis Doster, III,
James K. Page, Jr., Russell Bourne
Picture Editor: Nancy Strader
Assistant Editor: Amy Donovan
Picture Research: Frances C. Rowsell
Production Manager: Ann L. Beasley
Graphics Coordinator: Patricia Upchurch
Copy Editor: Florence H. Blau
Production Assistant: June Armstrong
Business Manager: Thomas A. Hoffman
Assistants: Therese Gates, Sara Munson,
Jamie Shepherd
Marketing Representative: William H. Kelty

Design Direction: Phil Jordan,
Beveridge & Associates, Inc.
Mechanical Preparation: Midnight Oil
Separations: Lanman Lithoplate, Inc.
Typography: Custom Composition Company
Printing & Binding: Holladay-Tyler
Printing Corp.

Library of Congress Cataloging
in Publication Data

Fire of Life.

Includes index. 1. Sun. 2. Photobiology.
3. Solar energy. I. Smithsonian Institution.
QB521.F57   523.7   80-28422
ISBN: 0-89599-006-7

Contents

Introduction

Our Onlie Begetter

S. Dillon Ripley

It is entirely appropriate that the Smithsonian should bring you a book about the sun, the dominant body in the solar system. No other object in the universe has so stirred the concern of mankind throughout known history, as well as in prehistory. We know from the megalithic structures of Stone Age and later how carefully the appearance of the sun throughout the seasons of the year was followed and measured as a central phenomenon, long before the concept of time as we think of it today—the hours and minutes of our busy lives—impinged on human consciousness.

The sun too has played a prominent part in the research interests of the Smithsonian itself during the last 100 and more years. From the time of Joseph Henry, the first secretary of the Institution, to the present, each head of the Institution has had a greater or lesser part to play in stimulating solar research. Today, no less than in past years, the Smithsonian conducts studies on aspects of the sun's effect upon the Earth and all its life, as well as the physical phenomena that control life, with the hope of benefitting mankind.

Joseph Henry, feeling that the establishment of a physical observatory was a matter of highest priority for the United States, wrote to an

unnamed colleague in 1870, urging upon him the necessity of such an observatory:

We now know that the sun is undergoing remarkable changes, the character of which can only be ascertained by the results of accurate observations compared with those of experimental investigation. The observer should divide his attention between the phenomena revealed by a critical and continued examination of the sun and the production of similar phenomena in the laboratory. In this way European investigators . . . by the application of the spectroscope, different modifications of the telescope, and other lately invented appliances . . . have arrived at most interesting results.

Lion and sun crown the Smithsonian Mace, carried for academic and formal occasions by the Institution's chief executive officer, the Secretary. The heraldic demi-lion rampant holds a sun in splendor. The five-inch feline stands on an inset of smithsonite, a mineral identified in 1802 by James Smithson, the Institution's great benefactor. The Washington Monument in the United States capital is modeled on a solar cult structure of Egypt—the obelisk. To the monument's left, two Smithsonian towers rise near the site of the Institution's first solar observatory. As the sun moves, shadows of the towers and the monument move across the Mall in sundial fashion.

Ceremonial figures flank rainbow designs on a painted elkskin from the pueblo at Oraibi, Arizona. These kachinas represent spiritual messengers who carry the prayers of the Hopi people for good harvests. Below, a rainbow crosses arid Arizona flatland. With it comes the rain and the mill-spinning winds that raise hopes for better crops, brighter tomorrows. In some cultures the rainbow has been considered a harbinger of disaster, a notion contrary to the optimistic lore of the West's Judeo-Christian tradition.

Henry, perhaps concerned that the creation of a Smithsonian observatory would take too large a proportion of the available Institution funds, did not further the project. However his successor, Spencer F. Baird, took a long step forward by employing as his assistant secretary Samuel P. Langley, one of the fathers of modern aviation, a solar physicist, and an inventor of astronomical apparatus. He used a spectroscope to study gas emissions in the laboratory, correlating these with emissions from the sun's surface. Langley established an astrophysical observatory, made formally a part of the Smithsonian in 1890. In 1897 he stated:

It is now well understood that . . . every manifestation of life from that of the lowest vegetable form up through animal existence, to that of man, including all his works and industries, comes from the sun. . . .

Charles G. Abbot, a young assistant in 1896 when Langley experimented with one of his first heavier-than-air flying machines, eventually became the fifth secretary of the Institution, and concentrated his entire career, lasting well over 70 years, on studies of the sun and its influence upon the Earth. One of his principal concerns was the understanding of sunspots and the ebb and flow of radiation reaching Earth. His tables on the solar constant, measuring the energy rate received from the sun year in and year out, are the most comprehensive ever made. Under his aegis field observatories were established in deserts around the world, in our own Southwest, in Chile, South West Africa, and Mt. St. Katherine in the Sinai. In 1929 he established the division now called the Radiation Biology Laboratory, to study the correlation between plant growth, photosynthesis, and measured components of sunlight. Particularly important in recent years has been the Laboratory's work on ultraviolet emissions from the sun. There is now a growing realization of the vital applied value of such knowledge in connection with studies in meteorology, biology, medicine, and agriculture because of the perceived decline in the ozone layer in the Earth's atmosphere, which shields the Earth from harmful radiation.

Abbot retired as secretary in 1945, but continued his own research in solar radiation, writing tirelessly about the possibilities of solar energy for people's use. His last patent was an "Apparatus for Converting Solar Energy to Low

The fern Lindsaea ulei *spreads its light-trapping leaves near Manaus in Brazil's vast Amazonia. People here, and in other jungle regions, have recently initiated massive deforestation to help revitalize their economies in the face of rising energy costs. Tropical forests, by virtue of their furious rate of photosynthetic activity, do indeed represent a treasure trove. Yet the green bounty of the tropics is more vulnerable than it looks, and should the delicate web of forest life unravel through abuse, life elsewhere on the planet could suffer. In particular, vital production of atmospheric oxygen from photosynthesis might decrease, and shifts in climate could pose potential danger to established agricultural patterns.*

Sun-seeking eye glows at dawn on Arizona's Mt. Hopkins. Part of a gamma-ray study by the Smithsonian now serves as a sun-powered collector for Smithsonian and University of Arizona.

Cost Power" granted in 1972 when he was 99 years old, a record in itself.

Leonard Carmichael, seventh secretary, was instrumental in initiating the collaboration between the Smithsonian and Harvard University in 1955, now titled the Center for Astrophysics, perhaps the largest organization of its kind devoted to astronomical research in the nation. With the University of Arizona, the Smithsonian operates the nation's third largest and only multi-mirror telescope, on Mt. Hopkins south of Tucson, dedicated in 1979.

Key research of the Center revolves around the Langley-Abbot Research Program, named for these pioneers in the study of long-term solar variability. This includes theoretical investigations of magnetic diffusion in the photosphere and observational analysis of the location and rate of emergence of magnetic flux elements revealed by x-ray bright points. More and more, information on sunspots and solar flares, known to influence the jet streams in the lower atmosphere, seems to be correlated with our understanding of climatic cycles.

What is the point of this amassing of esoteric information? Abbot always felt that such knowledge would lead to our understanding of how to harness the incredible energy of the sun for man's benefit and use. As we know, now the Earth intercepts less than one part in two billion of the energy of this medium-sized star, some 93,000,000 miles away from us, on which all life depends. But this fraction of energy, if utilized somehow, could answer all of mankind's energy needs!

Solar energy offers the potential of heating and cooling the homes of people everywhere, preserving foods, and running factories, as well as providing transportation. A whole new field of study involving plasma physics and studies of the interaction of gases and magnetic fields may lead to new ways of producing electrical power by controlling thermonuclear fusion. Philosophically speaking, there may be higher priorities in human understanding, but in terms of application to human life there is perhaps no higher applied scientific priority at the present time.

As one travels around the world it is fascinating to observe the contrasts in perception of the sun and its effect upon the Earth. In the highest latitudes, toward the poles, there is a surprisingly constant amount of solar radiation, not far removed in effect from that of the equatorial latitudes, due to the differential in the moist tropics provided by heavy cloud cover. The growing season at high latitudes is tightly concentrated, of course, to the time of light, the Arctic or Antarctic summer, when growth is sustained by intense sunlight and life processes capitalize on this spasm of illumination. In middle latitudes the rhythms are perceivable though not as violent, giving rise over ages of man's history to rituals of magic, worship, and the dominance of theocratically controlled tradition. With the advent of science as we know it, beginning with assumptions of time, followed by measured observation by Arab and European astrologers and mathematicians, and eventually with the invention of magnifying lenses and the telescope, it became possible to determine that the Earth was not the center of the universe. The foundations of belief in the Creation, in the order and measure of the universe were threatened in these gradual cosmic perceptions. What doubts this placed on the infallibility of mankind, our place in the cosmos, and the role of God, remain to be resolved.

More practically, I believe we should determine to understand the role that the sun can play in ameliorating the insistent destruction by mankind itself of its only home. If we can create a new system for harnessing the vast presence of the sun all about us, a source not possessed by any race or nation of mankind but common to all, how much better perhaps in the end will we understand the purposes for which life itself was created? We are still ignorant of many things, not least how best to live with each other, how best to keep from irretrievably fouling our own nests. In the intense struggle necessary to solve the secret of survival, there may yet reside truths which we can understand in time, truths of existence, perceptions of retained memory, of synthetic imagination, of intelligence itself. If such research can bring us to a better understanding of ourselves, then that should indeed suffice for the time being.

The chapters in this book range widely, for all of observed life is bound up in the single fact of the existence of the sun, our "onlie begetter." Everything that happens, from rainbows to the fall of sparrows, from snow to the winds that let us sail or fly, from the appearance of civilizations to their demise in the sands of time—all is dependent on our star. The paper on which this book is printed, the light by which it is read, and the hand which writes these words are all evidence of the "fire of life."

Charles G. Abbot, above, helped guide sun investigation during more than 75 years at the Institution.

17

S tonehenge, that shattered ring of gaunt gray stones situated
on bleak Salisbury Plain in southwest England, has
attracted the attention of travelers and scholars for centuries. When we think of ancient man and the sun, the popular
theory that this ruin was once an astronomical observatory—
perhaps even a "computer" for predicting eclipses—springs
to mind.

The observatory theory, however, leaves many questions
unanswered. For what purpose would prehistoric people have
observed the sun? Did they rely on it as a practical timekeeper,
as important for their everyday lives as clocks and printed calendars are for us? Was Stonehenge the center of a religious cult—
the setting for bloody sacrifices, perhaps—dedicated to an all-
powerful sun god? Or dare we imagine that the thinking of the
Stonehenge builders anticipated the cool speculations of today's

Shooting the Sun

astrophysicists, the product less of fear and superstition than of
Stone Age philosophy and science?

Archeological writer Evan Hadingham shares with us his
scholarly rambles through Stonehenge and other prehistoric sites
as he traces the roots of today's sun sciences—astronomy and
astrophysics. Smithsonian's Owen Gingerich links the classical
world's empirical scholarship to the Enlightment's final over-
throw of outworn scientific dogma. Smithsonian historian Tom
D. Crouch outlines the Institution's own initiatives in both solar
astrophysics and practical application of sun energy. Third
Smithsonian secretary and astrophysical innovator, Samuel P.
Langley, established the international unit for solar radiation,
the langley. He also wrote in 1889 of the "rude, enormous
monoliths of Stonehenge" that might represent an infancy of
astronomy, of observatory practice. The final contributor to this
introductory section of *Fire of Life*, William Livingston, employs
the great McMath solar telescope at Kitt Peak—the National
Observatory—located in Arizona. He reveals the technology
behind instruments that not only supply images of the sun's
surface activity, but also probe the great orb's inner workings.

19

Evan Hadingham

Ancient Man and the Sun

Now that Stonehenge lies in ruins, it doesn't speak straightforwardly to us. Like a cathedral, Stonehenge was built and rebuilt on the same spot several times over a span of at least seven centuries, starting around 2800 B.C. Successive construction work in earth, wood, and stone destroyed much evidence that would allow us to trace its presumed astronomical alignments with confidence. Indeed, confusion greets the casual visitor. Many stones still stand proudly in place where they were originally erected about four thousand years ago; many more are strewn in chaos on the ground.

Stand at the center of the ring and look to the northeast. Through the middle of the three great arches you will see the Heel Stone about 250 feet away. Here is the famous spot where the sun appears to rise on the longest day of the year, an event that draws curious onlookers by the hundreds to view the midsummer dawn. The association of this major axis of Stonehenge with the summer solstice was known at least 250 years ago, for it was in an account published in 1740 that British antiquarian William Stukeley first noted that the entrance faced northeast, "whereabouts the sun rises, when the days are longest." In 1965 astronomer Gerald S. Hawkins revived interest in this and other sightlines when he published his celebrated book *Stonehenge Decoded.*

There are several curious facts about this most famous of all prehistoric astronomical alignments. For example, today the sun does *not* rise directly in the path of the Heel Stone, but a little distance to the left, or north, of it. Furthermore, because of an extremely gradual shift in the Earth's axis over the centuries, the sun would have risen even farther to the left back in 2000 B.C. A few minutes after the first flash of light the sun would indeed have appeared right over the top of the Heel Stone, but by that time it would also have cleared the horizon by a distance equal to more than its own diameter. It would have been easy to line up either the first gleam of the sun or the full disk with greater accuracy than this, if indeed accuracy was what the builders were striving for. This point suggests that in about 2000 B.C., when the grandest of all the Stonehenge structures was raised, precision was not uppermost in the minds of the builders.

It requires an effort of imagination, a step outside our 20th-century preoccupation with hour-to-hour timekeeping, to understand the hearts and minds of prehistoric sun watchers. Significantly, when their descendants eventually began to create art works from about 35,000 B.C. onwards, their beautiful engravings and cave paintings already express a keen observation of specific seasonal events. We find depictions of the appearance of salmon at spawning time, of new-born fawns, and of stags in their autumn rut.

Cave art reached its magnificent climax at the sites that today bear such famous names as Lascaux, Niaux, and Altamira. During this period archeologists can show that Ice Age hunters

were regularly following the movements of such migratory animals as reindeer and salmon. Predicting the behavior of these far-ranging creatures meant the difference between starvation or survival.

The earliest unwritten "calendars" were inseparable from a knowledge of the growth of vegetation and the activities of animals long before the sun itself was systematically watched.

The next step towards a primitive timekeeping system was to compare the changes created by the sun with the rhythm of the moon. As late as the last century, many native communities—from the Koryak of Siberia to the Dakota of the North American Plains—still followed the year by observing the phases of the moon and calling each month after a seasonal event: "the month of thawing snows," "the ripening of berries month," "the molting of reindeer month," and so on. According to a gifted American researcher, Alexander Marshack, Ice Age hunters may also have recorded time in this way. Alongside certain animal engravings one finds series of abstract lines and notches that may represent counts of the lunar phases. A rough-and-ready matching of the cycles of the sun and those of the moon was a basic part of man's increasing control over his surroundings.

Problems loomed for any ancient astronomer who tried to refine such a calendar by watching and counting on a truly accurate, regular basis. An exact number of lunar months cannot be

fitted into the solar year. If 12 months are chosen they fall short of the actual length of the year by about 11 days. If, on the other hand, 13 is the number selected, the year is too long by about 19 days. Some communities avoided the problem altogether simply by recognizing a "blank" season when no moons were counted, as did the Kwakiutl and Bella Coola Indians of the Pacific Northwest Coast. By contrast, we know that among other groups such as the Plains Pawnee and the Californian Yurok there were lively, sometimes angry, debates over which month was which and whether the year contained 12 or 13 moons. Clearly the development of a truly refined calendar required an unusual intellectual leap forward, stimulated by the special conditions that also gave rise to such extraordinary monuments as Stonehenge.

There was one simple way for ancient astronomers to divide the year accurately and yet avoid the complication of fitting in the moon and its phases. The skies of southwestern North America are often brilliantly clear at sunset, while eroded pinnacles and far-off snowy peaks form striking outlines around the desert horizons. These surroundings may have encouraged regular sun watching as a tradition still practiced a few decades ago by the Hopi and Zuni tribes of the Southwest. By memorizing in detail the various horizon points where the sun set on its course throughout the year, the sun watcher could keep an accurate

solar calendar, with some dates fixed a mere four or five days apart.

There were urgent practical reasons for this preoccupation with the sun. Growing staples like corn and beans in the Arizona desert has always been a risky business. The average rainfall of 10–13 inches in the Hopi region is too low to support crops without special methods of flood-farming and irrigation. Furthermore the frost-free period during the summer months is barely longer than the growing season of the maize plant itself. Accurate sowing dates, together with certain plantings staggered at different intervals to avoid the risk of a massive failure, were essential to survive in this precarious landscape. Once the sun sank beneath the critical spot on the horizon the sun watcher would announce the time for planting like a town crier, or, as in certain Hopi villages, merely inform the heads of the principal families when it was their turn to set out for their fields scattered throughout the arid wilderness. Abundant prehistoric remains in the Southwest, including the ruins of actual observatories, suggest that these practices may have originated at least a thousand years ago.

There is a risk in assuming that ancient astronomy always has to make sense in terms of our own logic. However practical a part the horizon calendars played in the lives of the Hopi, they would not have separated sun watching in their minds from the fertility magic that was also vital to them. For example, the office of sun

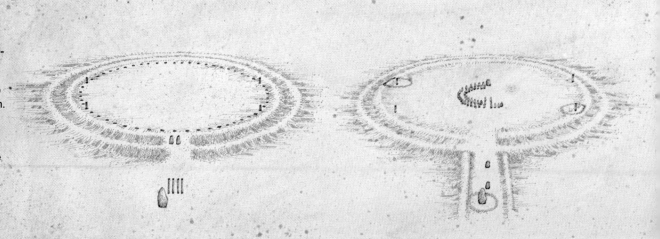

Successive generations of prehistoric Britons merged their efforts to create Stonehenge, termed a megalithic cathedral by some observers. Archeologists recognize at least four different phases of construction at the site on Salisbury Plain, beginning about 2800 B.C. and continuing over a millennium. First, a bank quarried from a chalk ditch outlined the perimeter of the monument. Just inside, 56 holes were dug and filled in almost at once, some

watcher carried with it obligations of good conduct, including sexual abstinence, to appease the sun. Sacred dances were no less essential to the well-being of the community than precise planting dates. We can appreciate the practical outcome of this system of sun watching, but the outlook and impulses that inspired it are unfamiliar to us.

When we return to the mysterious stone monuments of prehistoric Europe, we encounter problems never faced by those interested in native Americans. No written numerals, fragments of books, or carved hieroglyphs exist to aid us. Eyewitness accounts of ceremonies or farming practices are absent. When all we have are arrangements of stones and excavated pots and bones, conjecture takes the place of certainty. In this vacuum of understanding, the examples taken from native America may help us to ask the right questions. Did the builders of the ancient stone monuments devise an accurate calendar at least partly for practical ends, like the Hopi?

Recently the notion that the ancient Europeans developed advanced geometrical and astronomical skills has aroused great interest. A Scottish professor of engineering, Alexander Thom, has devoted 40 years to the study not only of Stonehenge but of hundreds of other stone rings and avenues built on a less awesome scale throughout the British Isles and Brittany. If his theories are correct, then the builders lined up the sun and the moon against distant horizon features as the Hopi did to make specific astronomical observations. At certain prehistoric sites, Thom claims, the exact day of midsummer or midwinter could be found, while other ruins represent the remnants of "a scientific study of the moon's motion." The power to predict eclipses is proposed as the motive that inspired elite groups of astronomer priests, whose achievements may certainly have matched those of the Maya Indians of Central America.

Thom's ideas have aroused a good deal of controversy, and a glance at only one of his "observatories" indicates the problems involved in such remarkable claims. At a place called Ballochroy, a desolate shoulder of moorland on the southwest coast of Scotland, three large slabs of stone stand in a line. The deliberate positioning of these stones is emphasized by the remains of a burial monument, a massive stone box (now empty), situated exactly on the same line over 100 feet away to the southwest. This is also the direction of the sun as it sets at midwinter behind the distant outline of an island on the horizon. Thom proposes that the ancient observers lined up the last flash of the sun with this horizon feature in order to detect the exact day when the sun "stood still" at midwinter. However, no evidence exists to confirm that they actually did this.

Indeed, a sketch of Ballochroy dating to the turn of the 17th century shows

Druids of old converge on Stonehenge, most famous of megalithic landmarks, in a fanciful scene from an 1815 aquatint. In fact, Druids had nothing to do with the building of Stonehenge.

containing burials. Arrangements of stones and posts suggest that ingenious lunar observations were already in progress. In the next phase, bluestones were removed, then replaced by a ring of tall sarsen stones linked by lintel slabs. Five huge archways rose at the center, followed by various rearrangements of bluestones within. During its final phases, Stonehenge seems to have been a symbol of astronomical lore, not a practical observatory.

that the stone box was then covered by a large cairn of stones, since destroyed, that probably blocked the view of the horizon altogether. While the layout of the site to the midwinter sun was surely deliberate, the builders may have been unconcerned with precise astronomy. All that was intended, perhaps, was that the sun should be seen to descend "into the tomb of the ancestors," and we can guess that this may have symbolized for them the journey of the dead to another world.

The link between death and the sun is clear at many other prehistoric sites in Britain. One of the most spectacular of them all is the mighty tomb of Newgrange in eastern Ireland, a collective burial place erected in about 3300 B.C. and decorated with magnificent

Ancient eyes—popularly believed to be those of a Great Goddess—gaze from a pot fragment discovered in a Stone Age passage grave in Denmark. Some kind of deity was probably worshipped by the tomb builders, but we have no name for any specific god or goddess. Light, however, seemed to have played a symbolic role in religious activities. Beliefs about death and the afterlife were probably linked to the sun at Newgrange, opposite, where at midwinter dawn the rays strike down the passage to the chamber.

carvings of spirals and other abstract designs. Underneath a gigantic mound of stone, clay, and turf, faced with dazzling, white quartz pebbles, a narrow passage twists its way to the central burial chamber. A slot above the entrance to the tomb admits a narrow shaft of sunlight shortly after dawn at midwinter. For a few theatrical minutes, this ray penetrates to the far end of the burial chamber nearly 70 feet away and lights up the stone basin placed there to receive the cremated remains of the dead. The entire design of the sealed tomb appears to have been arranged around this striking alignment with the sun. Once again, it seems that a powerful symbolism connected the dead with midwinter. In fact, persistent customs of orienting the dead can be traced throughout early British prehistory, and these strongly suggest to us an awareness of the sun's movement and its link with funeral ceremonies.

If the religious overtones of sun watching in prehistoric Britain are clear, we are almost in the dark when it comes to the question of a practical calendar. A chaotic mass of traditional folklore survives in Britain, accumulated not only from prehistoric times but also from the later and very different worlds of medieval and successive periods. Many of these British folk customs relate to the May Day and November festivals that were directly linked to the timing of harvests and stock selection. Attempts to trace such celebrations to remote prehistory, or to reveal practical horizon calendars like those of the Hopi, have not been particularly convincing. The significance of midsummer and midwinter to the stone circle builders seems certain, but we do not know how they may have related these solar events to their agricultural routine.

Indeed, many years of research may be necessary before we can assess the true nature of the astronomy practiced at Stonehenge and other stone circles. The crudity of the midsummer sunrise alignment suggests that the final impressive temple erected at Stone-

henge in about 2000 B.C. was not a precise observing instrument. However, traces of the first features of the site more than 500 years earlier indicate that then real efforts were made to achieve accuracy.

In 1966 the public parking lot northwest of Stonehenge was extended, and during this operation three huge postholes dug into the chalk bedrock were uncovered. The timbers once placed in these holes may have been as tall as 30 feet, and would have broken the horizon when viewed from the monument nearly a fifth of a mile away. If the observer stood beside various features of Stonehenge believed to date from its early phases, significant settings of the sun and moon would have occurred in line with these posts, and observations of considerable accuracy could have been made. Here is intriguing evidence that at one stage the Stonehenge builders were striving for precision. The sun and moon cycles that they were watching, though, would not have benefitted the farmer in any obvious way.

When the great temple was eventually raised at Stonehenge, it may thus have been a *symbol* of celestial knowledge acquired over many centuries rather than a scientific instrument. The final layout may even commemorate proficiency in adjusting a calendar; if we count the number of stones and pits, the totals suggest that a way had been found of reconciling the length of the lunar months and the solar year. This seems to have been achieved only by the Stonehenge people, since the builders of most other stone circles did not choose astronomically significant numbers of stones when setting out the monuments (even though some of these lesser rings seem to include sightlines to the sun and moon). Again, this emphasizes the unique character of Stonehenge and leads us to wonder how far an elite group of astronomers supported at this special place could have advanced in their knowledge of the sky.

Certainly a counting system must have existed in ancient Britain, but so far no trace of written numerals, a sys-

tem of dates or monumental inscriptions has yet come to light. It is difficult to believe that oral records alone could have been used to detect the complicated time patterns of recurring eclipses. It is just conceivable that the connection between eclipses and the orbits of the sun and moon could have been spotted by direct observations over a fairly short period, yet this would have required an extraordinary flash of insight from a prehistoric Einstein. No doubt men and women of genius existed in ancient times, yet the evidence that eclipses were actually predicted is far from satisfactory.

Reviewing the theories of Professor Thom, commentators have been tempted to compare British sky watchers to the astronomical traditions that flourished in ancient Central America. Among the most significant were those connected with the two days in the year when the sun shines directly overhead at noon and casts no shadows. (This is known as the "zenith passage" of the sun and can occur only within the tropics.) The remains of ancient observatories suggest that attempts were made to fix with precision the exact days when shadows vanished. At Monte Alban in western Mexico, a stone shaft pointing vertically up at the heavens can still be used to frame the sun's disk exactly at noon on the days of zenith passage.

There can be little doubt that at least one of the zenith passage dates was linked to the practical round of agriculture. In some parts of Central America, native folklore still associates a shadowless day with the onset of the life-giving rainy season and the commencement of the agricultural year. Moreover, studies undertaken in the mountains of Guatemala more than 40 years ago indicate that there the descendants of the ancient Maya were timing the dates of crop planting by watching the sun at the horizon.

Nevertheless the stone inscriptions and bark books of the classic Mayan civilization tell us that they reached a much more sophisticated level of astronomical expertise. Many centuries before the Spanish conquest an elite group of astronomers made calculations of little direct value to the ordinary farmer. In their observatories they sighted the sun, the moon, and other celestial bodies through small openings in the walls or by means of crossed sticks which directed their gaze. With these simple aids the Maya developed a correction between their calendar year of 365 days and the true solar year, that was more accurate than our own leap-year correction. They also acquired a store of knowledge about eclipses and planetary movements, and managed to measure the length of lunation—period between new moons—to within seven minutes of our modern value.

The passion that drove Mayan astronomers to such exactitude was not fueled by disinterested scientific curiosity. Instead, complex religious beliefs stirred in them a profound anxiety about the passing of time. Their intricate calculations were directed toward astrological ends, notably the determining of auspicious and unlucky days, far removed from science as we define it. These persistent ancient beliefs were reflected in the everyday schedule of the peasant farmer, for it is said that the modern Maya work fast to complete planting on a favorable day so as not to overlap with an unlucky one. "Practical" watching of the sun by the Mayan farmer co-existed with the strange astrological system handed down by his gifted ancestors.

Returning to Britain, we *can* be confident that the sun was venerated, perhaps feared there 4000 years ago, and that there was an awareness of its movements. Beyond such general conclusions, the motivation of the prehistoric astronomers eludes us. The example of the Maya and of other native astronomers warns us that remarkable skills were often devoted to religious ends that seem bizarre and baffling to us. The midsummer sunrise at Stonehenge inspires in us emotions of awe, and surely these were experienced, too, by the original builders. The rest of their thoughts and feelings are lost forever.

Pre-Columbian Americans venerated the sun and studied the stars. From their observations, the Mayan astronomers invented an elaborate calendar system. The tall tower at Palenque, below, a Mayan palace of the 7th century A.D., may have served as an observatory. From an earlier era, Mexican craftsmen of the Teotihuacán empire sculptured a dual symbol of death and sun, right. Calendrical and mythical symbols mingle in the famous "Stone of the Sun." The design appears here in gold, on a recently struck Mexican coin, left. Following page: stone pattern of radiating lines crests high ground at Sacsahuaman—fortress of Cuzco, Inca city of the sun. Near the site, modern Peruvians stage an annual solar pageant based on lore from the old civilization.

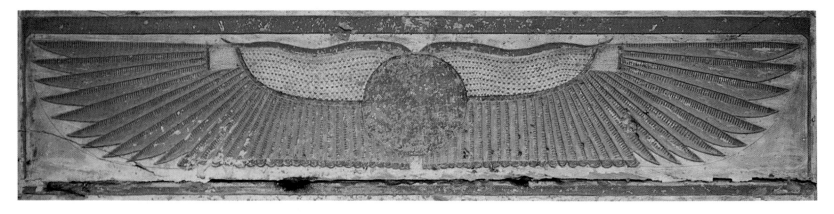

Great Revelations

Oh dear God, this is hell.
It's either dark
or I am dead.
God, I want to live!

Owen Gingerich

So exclaimed a terrified TV photographer on the slopes of Mt. St. Helens as the giant volcanic cloud obliterated the sun. For miles around, cities and towns experienced midnight at noon as the sun's rays were blotted into obscurity.

Fortunately for mankind the life-giving solar rays are so seldom extinguished that we take for granted our daily quota of sunlight. But throughout long ages past, there have been repeated moments of trauma as the invisible moon has nibbled away the solar disk. During most of the history of *Homo sapiens*, there has been so little understanding of the causes of eclipses that early man could only tremble in the horrifying expectation that the sun was about to vanish forever.

The searing fear that the sun might disappear had an alternative positive corollary: the very presence of the sun sustains all life on Earth and therefore that bright luminary was itself worthy of adoration. Around 1360 B.C. the Egyptian Pharaoh Akhenaten summed it up in his eloquent hymn to the sun:

Beautiful is your rising in the horizon of heaven, living Sun, you who were first at the beginning of things. You shine in the horizon of the East, you fill every land with your beauty. You are beautiful and great and shining. Your rays embrace the lands to the limits of all that you have made. You are far, but your rays are on the earth. . . .

The beings of earth are formed under your hand as you have wanted them. You rise and they live. Their eyes look at your beauty until you set and all work comes to a stop as you set in the West.

An echo, in a different setting, appears a few centuries later in the work of the Psalmist:

The heavens declare the glory of God; and the firmament sheweth his handiwork . . . In them hath he set a tabernacle for the sun, which is as a bridegroom coming out of his chamber, and rejoiceth as a strong man to run a race. His going forth is from the end of the heaven and his circuit until the ends of it and there is nothing hid from the heat thereof.

And in yet another context Nicholas Copernicus rises from his rather dense technical prose to praise the sun:

At rest, however, in the middle of everything is the sun. For in this most beautiful temple, who would place this lamp in another or better position than that from which it can light up the whole thing at the same time? For, the sun is not inappropriately called by some people the lantern of the universe, its mind by others, and its ruler by still others. Thus indeed, as though seated on a royal throne, the sun governs the family of planets revolving around it.

Despite our far greater scientific comprehension of the sun today, these poetic insights still hold. As a gigantic nuclear furnace, the sun provides, directly or indirectly, almost all of the energy used on Earth. Without its light and heat, life would quickly vanish from our planet. But the kinship runs even deeper. We are made of the same atoms as the sun—indeed, we are all recycled cosmic materials.

This, then, is our theme: how astronomers

over the past five centuries have built up an increasingly rich understanding of the sun and of our place among the stars.

The sun-centered or heliocentric concept of the solar system is now so commonplace that we forget the idea was unknown to ancient man and not generally accepted until the century of Descartes and Newton. It required a great leap of the imagination, a leap that was finally pressed into modern thought by the Polish cleric Copernicus in 1543.

The young Copernicus was an undergraduate at the University of Cracow when Columbus discovered America. Like all educated men of his day, Copernicus was taught that the world was round and firmly planted in the middle of things. The heavy terrestrial elements must pull themselves into a sphere, Aristotle had taught, and today we realize that his principle is even more widely true than he had imagined. There is no material strong enough to build a stationary non-spherical structure larger than a few hundred miles in extent.

Aristotle's physics was one of abstraction and generalization, but not nearly as removed from mundane reality as the physics of the 17th century was to become. To Aristotle, terrestrial motions continued only as long as forces were present, but in the heavens circular motions occurred eternally and naturally. Hence the entire commonsensical pattern of Aristotelian physics reinforced the belief that the world was unique and at the center of everything.

"It may seem to some people," the astronomer Ptolemy wrote in the second century A.D., "that there is nothing against supposing that the heavens are fixed and the earth turns upon an axis once a day." Indeed, such a thing might even appear to be simpler, but it is so physically unnatural as to be ridiculous, declared Ptolemy. Even Copernicus said that when he began to meditate on the mobility of the Earth, some 1400 years later, the idea seemed at first absurd.

Apart from the absurdity, there was nothing to prevent the arrangement from being considered as a strictly mathematical hypothesis; that is, the Earth could be spinning about its axis once a day rather than having the entire panoply of the heavens in swift diurnal rotation. The next mathematical step would be to let the Earth cycle about the sun once a year rather than vice versa. Such a hypothesis was put forward even in antiquity by Aristarchus of Samos, but it apparently got so little serious consideration from the ancients that we do not even know whether Aristarchus defended the arrangement as physically feasible.

What seized Copernicus' imagination, however, was not the simple relativity of motion between the Earth and sun, but rather his discovery that when the rearrangement was applied to the entire planetary system a remarkable economy and some beautiful linkages occurred. In the Earth-centered Ptolemaic system, each planet required two major circles to account for its observed motion. As Copernicus examined the planetary mechanisms, he realized that one of the two circles for each planet was always the same—that is, by appropriate geometrical transformations and scaling, each planet could have one of its two circles centered on the sun. Like a magician who can merge a series of separate metal rings into an interconnected set, Copernicus combined the planetary circles so that a single circle replaced five of the ten separate circles required for Mercury, Venus, Mars, Jupiter, and Saturn in the Ptolemaic scheme. In his *De revolutionibus* (Concerning the Revolutions) Copernicus wrote that "in this arrangement we find an admirable commensurability of the world and a sure harmonious connection between the motion and the size of each sphere." It was the Earth's orbit that provided a common measure, or commensurability, that established the entire scale of the system, fixing without ambiguity the relative distances of all the other planets. Moreover, the heliocentric system placed Mercury—the fastest planet—closest to the sun, and Saturn—the slowest planet—at the greatest distance, with the Earth falling in its harmonious place between the extremes.

Notice that Copernicus concocted his unorthodox arrangement strictly as a mental exercise. There were no new data that demanded this revision. Quite the contrary: not only did his theory lack observational evidence, but it required his contemporaries to suspend their beliefs in the traditional physics of earth, water, air, and fire. As Galileo later remarked, "I can never sufficiently admire those who have accepted the heliocentric opinion as true despite its violence to their own senses and what sensible experience plainly showed them to the contrary."

Hence, one cannot really call Copernicus the "discoverer" of the heliocentric system. Instead, he is its inventor, or perhaps more appropriately, its architect.

Winged disk symbolizes Egypt's dependence not only upon the sun but upon absolute monarchy and a mighty pantheon of nature gods. Akhenaten, whose solar hymn appears here, tried to change history. With his wife Nefertiti, the "Heretic Pharaoh" founded a gentle faith with a single and supreme god—the sun disk—represented above with hand-tipped rays.

Stars and zodiacal animals decorate the frontispiece from an early European translation of the Almagestum Ptolemei— one of the first printed books, and one that greatly influenced Renaissance scholarship. Ancient Egypt's great astronomical compendium and other scientific works from Greek and Roman times, existed only in manuscript copies and lay hidden for centuries in Islamic cities like Constantinople and medieval Cairo, right. Nicholas Copernicus began the demolition of the old Ptolemaic earthcentered universe when, in 1543, he published his De revolutionibus, opposite. Above right, an English version of the suncentered Copernican system, by Thomas Digges in 1576.

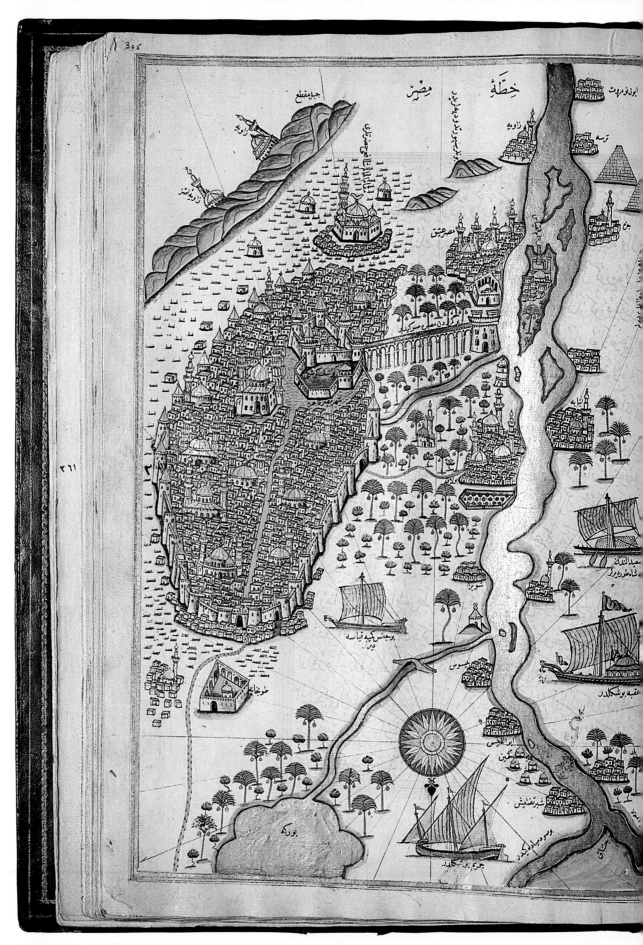

In the geometrical form that Copernicus left it, the heliocentric system could hardly be considered a physically real model for the universe. Nevertheless, his linkages between the sun and the planets were so convincing that the special role for the sun could not be abandoned. Many astronomers adopted the compromise system proposed by the Danish astronomer Tycho Brahe, in which the Earth remained implacably at rest while the swiftly orbiting sun carried the other planets around it in turn.

Such half-hearted measures were not enough for the young German theorist Johannes Kepler. Kepler demanded not only a geometry but a physics. What kind of mechanism or force could propel the planets about the sun? Groping for a force that could act across empty space, Kepler proposed a celestial magnetism. Ultimately unsuccessful, his theories and researches nevertheless gave compelling new reasons for believing that the sun was physically central in the planetary system. Copernicus had placed the sun at rest, but not in the actual center of the circular orbits. Kepler showed that the planetary orbits were elliptical and that the sun was always at a focus of each planetary ellipse.

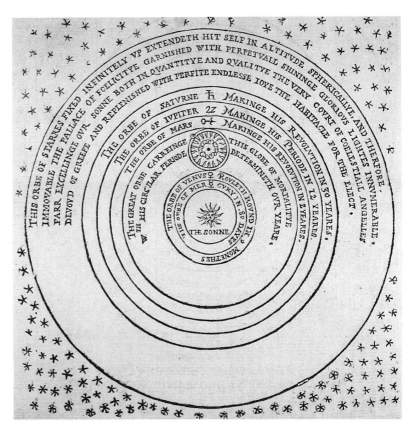

But it was Isaac Newton who, by taking physics to a higher level of abstraction, succeeded where Kepler had failed. Aristotelian physics was rooted in an everyday world where friction predominates. Newton imagined a world without friction and then showed that celestial motions closely approximated his idealized universe. In some ways, his third law of motion was the most imaginative leap of all: for every action there is an equal and opposite reaction. Applied to the Earth and sun, the third law meant that the force of the sun upon the Earth was exactly counterbalanced by the force of the Earth upon the sun. But because the sun is incredibly more massive than the Earth, the solar force bends the Earth's motion into its annual trajectory about the sun, while the equal force from the Earth swings the sun in an orbit a mere 600 miles in diameter, so small that for practical purposes the sun can be regarded as fixed. What is more, Newton's laws provided a convenient method for "weighing" the sun, showing that its mass is 300,000 times the Earth's. Following the publication of Newton's *Principia* in 1687, there could no longer be any doubt that the Earth was a planet orbiting the distant and massive sun.

33

In his *Rules of Reasoning in Philosophy* Newton stated that "to the same natural effects we must, as far as possible, assign the same causes," and among other examples he mentioned the similarity of "the light of our culinary fire and of the sun." From the image of the sun as fiery ball to the glorious vision of stars as suns seems a logical step by Newton's rule, and indeed this grand conception of the cosmos was increasingly accepted in the 17th century.

The ancient Greeks assumed that the sun, stars, and planets were all made of the same material, that is, of transparent and weightless "aether," but there the similarity ended. Aristotle, and even Copernicus or Kepler, would have been amazed to learn that the stars were suns, for this would imply that the stars were farther away than they could ever have dreamed possible. For them, the stars were all equidistant, fastened on the thin shell of the celestial sphere.

That the stars might be scattered through vast reaches of space rather than pinned on a spherical frame occurred to an Englishman, Thomas Digges, in the 1570s, and there are hints about the same time in the writings of the itinerant ex-friar Giordano Bruno. But the idea caught on slowly, and even the most avant-garde thinkers failed to adopt it until well into the 17th century.

Once the powerful concept of stars as suns became accepted, then it became clear that the constellations were more than the two-dimensional backdrop against which the planets moved. The fainter stars were surely farther away, giving hints of a grander universe than hitherto envisioned. Thus by the late 1600s a handful of leading natural philosophers saw a novel and challenging route of inquiry. If the stars were suns, then their distances could be calculated from the well-known rule for the diminution of light. When the distance of a luminous object is doubled, its apparent brightness diminishes by four; when the distance is tripled, the apparent brightness drops by a factor of nine, etc. Now if a star, Sirius for example, were a duplicate sun, and if its brightness were a millionth less, then it would be a thousand times farther than the sun (since a thousand squared is a million). But how could two objects of such vastly differing brightness be compared? This practical problem was first resolved by James Gregory, who in 1668 showed how to use the planet Jupiter as an intermediary mirror reflecting a known fraction

Suggesting the scholarly glow of the Enlightenment, Joseph Wright's "The Orrery," above, depicts a model of the solar system in use. A central candle represents the sun. Opposite: the reflecting telescope invented by Isaac Newton in 1671. His Principia, a milestone in the literature of science, included notes on the nature of gravity and laws of motion that govern orbits of planets and moons.

of sunlight. Isaac Newton used essentially the same method, with Saturn instead of Jupiter, although his calculations were published only posthumously. The Dutch savant Christiaan Huygens took a different tack by first observing the sun through a tiny hole, then reducing its light still further by means of a lens until its light seemed comparable to that of Sirius—surely a remarkable feat of perception to recall by day the brightness of Sirius the previous night! Gregory's calculation put Sirius at 83,000 times the distance of the sun, Newton's at 1,000,000 and Huygens' at 28,000.

Science moves ahead by distilling from the rich variety of nature basic underlying principles. Part of the strategy is the making of simplifying assumptions, such as motion without friction or stars as similar as peas in a pod. Yet such simplifications are potentially treacherous, for nature is often richer and more variegated than our assumptions. And so it is with the stars. Few are duplicates of the sun. Most are fainter, but the brilliant ones punctuating the night sky and composing the familiar constellations are generally intrinsically brighter than the sun. Thus as it happens Sirius is about 100 times more luminous than our daystar, so its actual distance is about 500,000 times greater than the sun's. But the attempts of the

17th-century astronomers should be seen not as a failure of result but as a triumph of conception. Though their numbers erred, they established the essential grandeur of the starry frame. In a single stroke the universe had been expanded to a size inconceivable to the most brilliant thinkers a few generations earlier.

After the magnificence of this achievement, a century elapsed before anyone had both the vision and courage to apply a scale to the entire structure of the nighttime sky. Most professional astronomers were concerned with celestial mechanics, with working out the trajectories of comets and planets against the distant celestial tapestry. It was a self-taught amateur astronomer and telescope maker who saw farther and who decided to grapple, as he put it, with the construction of the heavens. Abandoning his successful musical career, William Herschel set out to count stars. He again assumed "faintness means farness," but the principle worked far better in the aggregate than when applied to single specimens. The results of his measurements revealed a vast disk-like ensemble of stars, much like a grindstone, with the sun near the center. And there the matter rested for another century.

Not until the first two decades of the 20th century did astronomers gain an accurate perception of the rich variety of stars. Through the researches of a Danish astronomer working in Holland, Ejnar Hertzsprung, and the Princeton astronomer Henry Norris Russell, scientists learned that most stars were approximately like the sun: some brighter and hotter, many fainter, cooler, and smaller. But a handful were spectacular, hundreds of thousands of times larger and more luminous than the sun; these became known as giants or supergiants.

One of Russell's students, Harlow Shapley, helped establish these dimensions and masses by studying several scores of eclipsing binary stars for his doctoral thesis. Along the way he discovered that the so-called cepheid variable stars, whose light variations somewhat resembled those of close eclipsing binary pairs, were single stars rather than binary systems. Subsequently he showed that these rare pulsating variables had to be numbered among the highly luminous supergiants. All this stood Shapley in good stead when he got a job at Mt. Wilson Observatory in Southern California. Working with the world's largest telescopes, first a 60-inch reflector and later with the 100-inch, Shapley began a study of globular clusters, huge aggregates of thousands of stars concentrated beside the richest regions of the summer Milky Way, in the constellations Sagittarius and Scorpius. He quickly realized that the brightest stars in these clusters were giants and supergiants, including some dozens of Cepheid variables. Once more the principle of "faintness means farness" came into play, but with greater sophistication than before, because Shapley realized that he was dealing with high-luminosity stars, not ordinary ones like the sun. Quickly he grasped the implication: the globular clusters were very far away, farther than almost any celestial objects ever fathomed.

More slowly, a second implication of these findings dawned on Shapley. Since the globular clusters were concentrated in a single direction in space, were they not perhaps distributed around some distant center of the Milky Way galaxy? Shapley thought so, and proposed the grand leap from heliocentric to galactocentric, with the hub of the universe perhaps 50,000 light-years away from the sun. Like Copernicus, Shapley had his doubters—after all, the statistical astronomy had invariably seemed to confirm Herschel's pivotal centrality for the sun. Not until the 1920s, when other astronomers proved that the sun was actually wheeling around a distant center as part of the giant Milky Way pinwheel, did the resistance cease. And not until 1930 did astronomers finally learn why the statistical procedures gave the wrong result: space within the Milky Way is peppered with interstellar dust, creating a subtle fog that closes off our vision and simulates a heliocentric galaxy.

It remained for Edwin Hubble, Shapley's great rival at Mt. Wilson, to apply the Cepheid variable yardstick to the faint pinwheel or spiral-shaped objects being captured in great abundance with the California telescopes. Once he found the cepheids in the Andromeda nebula and a few other spirals, the "faintness means farness" concept once more came into play, and Hubble established distances of around a million light-years for the brightest spirals. Proceeding on the same principle, but for spirals so faint that individual stars could no longer be distinguished, Hubble recognized them as other galaxies and described a universe whose size was reckoned in hundreds of millions or even billions of light-years.

Had Copernicus been able to catch just a glimmer of this ocean of space and yet maintain his sanity, he might have pointed to a curi-

ous anomaly. Though the sun was but an ordinary star in the fringes of a vast stellar system, nevertheless our home was the largest, most brilliant galaxy known. Not until 1952, when astronomers undertook a thorough recalibration that doubled the distance to the Andromeda galaxy and quadrupled its luminosity, was our Milky Way seen to be a large but not extraordinary galaxy.

Meanwhile, a century earlier, a German chemist and his young physicist colleague took a giant step forward by establishing the new discipline of cosmochemistry. "Kirchhoff and I are engaged in a common work that doesn't let us sleep . . ." So wrote Robert Bunsen to a friend in November 1859. Bunsen, whose name is familiar to anyone who has ever worked in a science lab, was a chemist searching for new elements. Gustav Kirchhoff was a physicist who had shown Bunsen how to use a prism to analyze the colors that resulted when various elements were placed in the flame of a Bunsen burner. They knew that each element produced a unique pattern of colors, the so-called bright-line spectrum. They also knew that the solar spectrum was a continuous rainbow, but with numerous fine lines separating the brighter

Radio telescopes, like the 250-foot dish at Jodrell Bank in England, reveal data hidden to optical instruments. A similar instrument, the 140-foot receiving antenna at Green Bank, West Virginia, caught signals from space that produced the image opposite, a view of the center of our galaxy where a powerful gravity anomaly —a Black Hole—may lie.

colored intervals. What Kirchhoff had unexpectedly discovered was a way to produce the dark lines of the solar spectrum artificially, by passing a continuous spectrum through a flame doped with one or more elements. This meant that they had found a way "to determine the composition of the sun and stars with the same accuracy as we determine sulfuric acid, chlorine, etc., with our chemical reagents."

Kirchhoff's spectral analysis unlocked the chemistry of the universe; it revealed that the sun, as well as the stars, was generally made of elements familiar on Earth: hydrogen, sodium, calcium, iron, chromium. One series of spectral lines was unidentifiable. Before the century was out this enigmatic solar element—helium—was discovered on Earth in natural gas wells.

Only after the structure of atoms was much better understood (in the early years of this century) did quantitative analysis of the sun become possible, and this eventually provided the key to another outstanding mystery, namely, the source of the sun's energy.

As long as the Earth's history was figured in a few thousand years, the sun could be likened to a culinary fire; after all, a chunk of anthracite coal the size of the sun would burn that long. But as geologists claimed an ever greater antiquity for our globe, finding a sufficient source for the sun's radiative output became a major desideratum for physicists. Lord Kelvin calculated that the sun could have slowly contracted for a hundred million years, gradually converting its gravitational potential energy into heat and light, but even this was insufficient to account for the time span of the geological record.

By 1920, Arthur Eddington argued with some verve that nuclear energy could power the sun. Very presciently, he wrote:

If, indeed, the sub-atomic energy in the stars is being freely used to maintain their great furnaces, it seems to bring a little nearer to fulfillment our dream of controlling this latent power for the well-being of the human race—or for its suicide.

Nevertheless, Eddington could give no details of the process for converting atoms into energy.

In fact, without a better knowledge of the sun's composition, there was no way to establish the nuclear reactions that could drive the sun. The relative abundances of the solar elements were first derived quantitatively by

Cecilia Payne, a young English scientist working at the Harvard Observatory, but her numbers were astonishing beyond belief. In the Earth, iron is the principal constituent, but her calculations showed that hydrogen was a million times more abundant than iron on the sun, with helium running a distant second. For several years no one, including Miss Payne herself, accepted these results until H. N. Russell and several others finally realized that a variety of astrophysical problems suddenly fell into place if they assumed that the sun differed enormously in composition from the Earth.

Once astronomers understood that the lightest element, hydrogen, was the principal constituent of the sun, it rapidly became obvious that hydrogen must serve as its nuclear fuel. A general outline for the fusion of hydrogen into helium promptly appeared, although the specific pathways were not established until further laboratory work had been done. Even though it is impossible to observe the solar interior directly, astrophysicists are fairly confident that they have deduced correctly the nuclear processes that have powered the sun for the past five billion years.

The night has a thousand eyes
And the day but one;
Yet the light of the bright world dies
With the dying sun.

Francis Boudillon's wonderful word imagery recalls the jungle and the fearsome dark despite being written for turn-of-the-century minds who knew full well that stars were suns. It forcefully reminds us of the overwhelming and essential power of our celestial hearth, the one star near enough to really matter.

But do not discount the other stars! For in their living and dying there are cosmic connections that touch us and our sun.

Although the sun's composition differs greatly from that of our planet, curiously enough, the composition of living matter follows not the Earth but the sun. Our bodies, like the sun, have more hydrogen atoms than any other. Helium, a noble gas unable to form compounds, is found abundantly in the sun though not in us; but then carbon, oxygen, and nitrogen occur in the same order as in that bright eye of day.

How is it that "brother sun" (as St. Francis called it) is so kindred to us? Even more billions of years ago, when the universe was created in a

fiery cataclysm, the initial bundle of energy transformed itself primarily into hydrogen and helium. There was neither sufficient iron for our hemoglobin nor magnesium for the centers of chlorophyll molecules. In the primeval generations of stars, a great elemental reprocessing occurred; what might rather vividly but somewhat inaccurately be called core meltdowns took place in the incredibly hot interiors of supernovae. These stars spewed out their cosmic ashes with explosive force, the hydrogen and helium having long since been fused into heavier atoms.

From these cosmic cinders the sun, the planetary system, and ourselves have obtained among others, the carbon, oxygen, potassium, and iron that we so vitally need. Without them we could not live, and without them the sun would long since have died, having lived an energetic and much shorter life.

We now understand much about the sun—indeed, perhaps too much, for we have stolen some of the secrets of its cosmic fire and have brought them to Earth. To paraphrase Eddington, will they be for our well-being or for our suicide?

The enigmatic core of the Milky Way galaxy seems even more mysterious in false-color enhancement. By using the computer-generated highlighting, scientists can gain an intellectual grasp of the hub of our galaxy. Curiously, outbreaks of solar magnetic force can resemble vast surges of magnetic energy that occur at the heart of some galaxies— in the region of so-called Black Holes.

S. P. Langley, Del.

The Solar Constant

Tom D. Crouch

A bold, radiant sunburst! How fitting that such a device should serve as the central element for the great seal of the Smithsonian Institution. For almost six decades, from 1886 to 1944, solar research was a dominant theme of Smithsonian science. During these years the Smithsonian Astrophysical Observatory (SAO) grew from a handful of specialists working under primitive conditions in a simple wooden shed to become an international center for the study of the sun.

From the outset, the staff of SAO had little interest in traditional telescopic observations of the stars and planets. Rather, they created a laboratory in which precision instruments were developed to probe the secrets of the sun's enormous power and its effects on the Earth.

At the Smithsonian, astrophysics was regarded as a "practical" science. Its practitioners were not theorists intent on uncovering the structure of the universe, but men who believed that their work held the promise of direct and immediate benefits for mankind. It was an age of Baconian science triumphant. To have real value, scientific research ought to be directed toward utilitarian goals, and what could be more useful than a better understanding of the way in which the sun, "the material ruler of our days," affected conditions on Earth. Perhaps "Uncle Joe" Cannon, the venerable Speaker of the U.S. House of Representatives, put it best during a House debate over an early SAO appropriation. "Everything hangs upon the sun, sir," he remarked to a skeptic, "and it ought to be investigated."

Smithsonian researchers agreed, and the direction of their studies remained clear and unchanged for 60 years. Their goals were simple and well defined. They sought to understand the nature of solar radiation, to measure accurately the amount of solar energy reaching the Earth, to understand the way in which the atmosphere transmitted and altered that energy, and to investigate the impact of solar radiation on the Earth.

The story of solar science at the Smithsonian really begins with the career of Samuel Pierpont Langley, named third secretary of the Institution in 1887, although first secretary Joseph Henry had written of the need to study the sun as early as 1870. One of the least understood figures in the history of 19th-century American science, Langley is remembered primarily as an aeronautical pioneer, the man responsible for a large man-carrying "aerodrome" that failed to fly when tested on two occasions in 1903. As a result, he is almost invariably portrayed as an unsuccessful engineer rather than the pillar of contemporary science that he was.

A native of Roxbury, Massachusetts, Langley was born on August 22, 1834, the son of a prosperous produce merchant. He rejected the opportunity for a Harvard education in favor of apprenticeship and a career as an architect and civil engineer in Chicago and St. Louis. Not much is known of Langley during these years, but we can assume that he was less than satisfied with his situation, for in 1864 he returned to Boston and his first love—astronomy.

As a scientist he was self-trained. His education consisted of a stint at amateur telescope making and a tour of European observatories. This experience was capped by a veneer of academic polish gained as an assistant at the Harvard Observatory and a few months on his own superintending the reestablishment of the astronomy program at the U.S. Naval Academy following the Civil War.

In 1866 he moved to the Western University of Pennsylvania where he served as professor of physics and director of the Allegheny Observatory. He remained in Pittsburgh for 20 years, creating a far-sighted, imaginatively conceived research program that transformed the observatory from a backwater haven of amateur astronomy into a world-renowned astrophysical laboratory.

It has long been assumed that Pittsburgh's growing clouds of industrial smoke and smog forced Langley to turn his instruments toward the brightest object in the heavens. In reality, his decision to undertake solar research was much more complex.

The sun had fascinated Langley since childhood. "I used to hold up my hands," he would later recall, "and wonder how the rays made them feel warm, and where the heat came from, and how." Langley's own belief that science should pursue useful goals, coupled with his taste and talent for experimental research, led him to a search for answers to the questions that had puzzled him as a child.

His first venture into solar science came in 1869, when he accompanied an expedition to observe a solar eclipse. Inspired by the "grandest spectacle nature offers" he decided to "study the physical constitution of the heavenly bodies, and especially that of the Sun."

Langley's early reputation was based on a series of sunspot drawings prepared after his return to Pittsburgh. He was initially convinced that sunspots influenced the Earth's climate "by decreasing the mean temperature at their maximum," but quickly concluded that much more solid evidence than his subjective visual observations of the solar disk would be required to prove such a direct relationship. Hard facts, precise data on both solar and meteorological variations would have to be obtained, analyzed, and compared.

Thus began Langley's search for the solar constant, the average amount of solar energy available above the atmosphere at the Earth's mean distance from the sun. Once this basic figure was determined, variations could be noted and perhaps linked to climatic changes. But what sort of instruments were available in the late 19th century to an experimenter who sought to measure minute fluctuations in solar radiation above the protective blanket of the atmosphere?

The pyrheliometer, a device that measured the total intensity of solar radiation falling on a receiver was another possibility. Yet the pyrheliometers of the period were too crude and insensitive to suit the perfectionist streak in Langley's character. Moreover, these instruments could only gauge the total energy received at ground level. The photometer, which Langley began to use in 1875, was one possibility. This instrument made it possible for an observer to establish a rough relationship between the sun's brightness and presumed temperature and that of a blast furnace or other luminous object on Earth. However, it could not provide the absolute quantitative measure Langley sought.

By 1878 Langley had finally developed an instrument to meet his needs. He called it a bolometer, or "ray measurer." Stripped to its

Passion for precision guided Samuel P. Langley's quest for solar knowledge, as reflected in his detailed illustration of a sunspot in 1873. A pioneer in the science of astrophysics, Langley went on to become the third secretary of the Smithsonian and to found the Astrophysical Observatory. Below, Langley views the solar eclipse of 1900.

essentials, the bolometer consisted of two thin platinum strips equal in dimension and electrical resistance. Painted black, the two strips absorbed 97 percent of the solar radiation falling on them. A storage battery was connected to the system, but as there was an equal electrical potential between the strips, no current flowed through the circuit under normal circumstances—both strips the same temperature.

To take a reading, one of the bolometer's platinum strips was masked while the other was exposed to the light of one portion of the spectrum. As the exposed strip became warmer, its resistance increased and a measurable current could then be detected by a sensitive galvanometer. Thus, the bolometer magnified minute changes in temperature, producing an effect, as Langley remarked, like a finger on the throttle of a steam engine.

By 1881 Langley's bolometric studies enabled him to calculate a value for the solar constant of 2.84 calories of heat per square centimeter per minute, a much higher figure than had been proposed by earlier experimenters. In addition, Langley concluded that the various spectra of the solar radiation differed only in wavelength and did not represent three different kinds of rays, as some had argued.

When he realized that his conclusions ran counter to general opinion, Langley became anxious about the possible impact of Pittsburgh's smog on his readings. In 1881, with the assistance of the U.S. Army and a local philanthropist, he led an expedition to California's Mt. Whitney, where he hoped to confirm his earlier observations. Assisted by James Keeler, a recent graduate of Johns Hopkins, Langley battled desert dust storms, mountain forest fires, and the hardships of travel by mule in order to operate his bolometer in the high, clear air of the Sierras. On the basis of the Mt. Whitney data, he felt confident in offering a new and even higher estimate for the solar constant of 3.0 calories per centimeter per minute. This figure would stand for more than 20 years, until finally corrected to 2.1 calories by another Langley assistant, Charles Greeley Abbot.

Another result of the spectrographic studies conducted on Mt. Whitney was the extension of the solar spectrum into the region of the invisible far-infrared wavelengths. The English astronomer William Herschel had discovered the existence of the infrared portion of the spectrum in 1800, and several studies of the region had been conducted over the ensuing years. But not until Langley combined a good diffraction grating (an instrument for separating with great exactness the lines of the spectrum), his bolometer, and the excellent experimental conditions on Mt. Whitney, was anyone able to study infrared radiation in detail.

However, by that time Langley's personal involvement in scientific research was coming to an end. By 1887 his work had brought him the highest academic and professional awards, including honorary degrees from such leading universities as Oxford, Cambridge, Harvard, Yale, Princeton, Wisconsin, and Michigan. When offered the post of assistant secretary of the Smithsonian, he eagerly accepted.

In November 1887, before Langley could reach the nation's capital, aging Secretary Spencer Fullerton Baird died. So, immediately after his arrival in Washington, Langley was named the Institution's third secretary. Soon after, he began the process of selecting a local site for an astrophysical observatory. Negotiations with Arlington Cemetery officials failed. An attempt to add the facility to the proposed National Zoo met with opposition from Congressmen already skeptical about the Smithsonian's venture into "bear farming." Langley then ordered the construction of a temporary observatory structure on the Mall, in the shadow of the "Castle" that served as administrative headquarters of the Institution.

The squat wooden building, originally painted a reddish brown (roof and all), was hardly an ideal laboratory. During the long, humid Washington summers the temperature inside often climbed to 120°F. One of Langley's assistants recalled that during his first summer in the building he was able to do little more than "lie on a table and sweat." Conditions became unbearable when the summer humidity grew so high that the rock salt prisms fogged, and a coal furnace was placed in operation to dry the air. Langley, who abhorred newsmen, allowed the grass to grow tall around the observatory, giving the place a neglected air that discouraged curious reporters.

The cold, clear air of the Sierras lured Samuel Langley to Mt. Whitney in 1881. Here he worked to document "the constant"—the sun's total output of radiant energy.

Nor did Langley's crusty impatience and insistence on perfection ease the lot of early SAO employees. When queried as to why the secretary had never married, one long-time colleague ventured, "I don't know . . . I suppose he wanted to be married today and raise the whole family tomorrow and no woman would undertake it."

Not until June 1895 did the secretary finally discover an assistant in whom he could place the greatest confidence. Touring the Massachusetts Institute of Technology, he was introduced to Charles Greeley Abbot, a 23-year-old master's candidate in physics. Much impressed, he offered the young man a position with the SAO at the princely salary of $1200 a year.

Abbot was an extraordinarily talented experimenter who surpassed even Langley in the search for ever more accurate instruments. Throughout the 1890s, as the secretary's interest focused on aeronautics and the problems of administration, Abbot gradually accepted leadership of the Observatory.

Under Abbot's guidance, SAO scientists continued Langley's earlier research into the infrared region with the bolometer. By 1891 photography had been introduced to the study. A clockwork mechanism was used to move the spectrum slowly across the instrument while a mirror traced the movement of the galvanometer needle on a photographic plate.

Abbot increased the accuracy of existing equipment and devised various means for isolating the instruments from vibration, temperature changes, and other conditions that introduced an element of error into the readings. As a result of this effort, he produced an improved bolometer that could be read "as easily as a mercury thermometer." An energy curve for the entire solar spectrum that had taken days for Langley and Keeler to record on Mt. Whitney could now be produced in a few minutes.

As instruments and experimental techniques were improved, Abbot grasped the practical potential of solar research and documentation. By 1903, he had fully adopted Langley's original goal of discovering a direct connection between solar radiation and climate.

The first real opportunity to investigate solar influences came in March 1903, when SAO observers noted a sudden drop of 10 percent in the solar constant. Langley and Abbot were elated when their search of daily weather records indicated a subsequent drop in temperatures in the Northern Hemisphere.

Abbot was dispatched as a member of two solar eclipse expeditions in order to carry on the search for solar variation under clearer skies. In 1905, the SAO established a solar observatory on the summit of Mt. Whitney, the first of several such installations Abbot would construct to gather additional evidence of solar variation.

Langley died of a series of strokes in 1906, just as the solar research was beginning to show real promise. As Abbot remarked, "his dreams of finding the solar causes of the good years and lean ones, like his dream of teaching men to fly, he lived to see supported, but not fully established, by the researches he initiated."

With the approval of Charles D. Walcott, the new secretary, Abbot continued to emphasize the development of improved instruments. With a firm value established for the solar constant and better understanding of the effect of atmospheric absorption on solar energy, he felt the time had come to supplant, at least partially, the use of the complicated bolometer with simpler, more portable instruments that required less training and skill to operate. New types of pyrheliometers were perfected, devices that gauged the total amount of solar energy reaching the Earth's surface. The pyranometer was added to the astrophysical tool kit to measure that portion of the sun's energy that is diffused by dust in the atmosphere. Other exotic instruments, including the melikeron, the kampometer, the periodometer, the slide-rule extrapolator, and highly sensitive radiometers made their appearances in due course. All were aimed at providing a simpler, quicker, and more foolproof method of acquiring and evaluating a detailed record of solar variations.

Abbot was also intrigued by the possibility of sending automatic recording instruments into the upper atmosphere. By 1914 the first of these, the balloon pyrheliometer, was being flown to altitudes of 15 miles.

Two years later, Abbot received a letter from Robert Hutchings Goddard, a professor of physics at Clark University. For some years he had been conducting experiments with liquid pro-

An electric wind charger tops dry Chilean Mt. Montezuma, site of a solar station established in 1920 by Charles G. Abbot, Langley's successor in sun research.

pellant rockets. His work had reached a point where additional funding would be required if progress was to continue.

Abbot recognized the potential of Goddard's project and was enthusiastic over the prospect of sending his instruments beyond the atmosphere. He carried the young physicist's case to Secretary Walcott, who approved an initial grant.

It was the beginning of a relationship that would last until Goddard's death in 1945. Despite the fact that Goddard was unable to develop a rocket powerful enough to carry a scientific payload, Abbot remained his most important friend in the scientific establishment. He published Goddard's two key reports on rocketry, funded work that included the development of the world's first liquid propellant rocket, and assisted him in locating funds.

But most of Abbot's energies during the years immediately prior to World War I were directed toward expanding the network of solar observation posts. This process began with the appearance of the first serious evidence in support of a direct connection between solar energy and weather. In 1915, H.H. Clayton, chief forecaster of the Argentine meteorological service, compared Abbot's record of solar variation with local weather with the hope of finding a more reliable means of making long-range weather forecasts. The results strongly suggested that a rough correlation between temperature, barometric pressure, and the rise and fall of solar energy levels did exist.

On the basis of these studies, Abbot felt justified in drawing on the Smithsonian's Hodgkin's Fund to establish a new solar observatory at Calama, Chile. Later moved to the more isolated Mt. Montezuma, some 12 miles south of the original site, the observatory afforded the ideal combination of clear skies and dry air that made possible readings accurate to a millionth of a degree. It was at Mt. Montezuma that a tunnel, or "cave laboratory," was first dug to house the delicate instruments. Each of the six major Smithsonian observatories established between

1915 and World War II, from Mt. Harqua Hala in Arizona to Mt. St. Katherine, only 10 miles from biblical Mt. Sinai in the Sinai Peninsula, would be built on the same tunnel pattern.

By 1919 SAO staff members were also refining the techniques used to gather data at each station. From this point until Abbot's retirement as fifth secretary of the Smithsonian Institution in 1944, the astrophysical research program settled into the routine of "highly difficult, often disappointing, day-to-day" rounds of the same observations made time and again.

Having devoted his career to the study of the energy radiating from the sun, it seemed only natural that Abbot should attempt to harness this enormous power source. With the assistance of John A. Roebling, a wealthy benefactor who had earlier provided financial support for the Smithsonian studies of solar radiation and weather, Abbot patented a variety of solar heaters and cookers or ovens. The first of these was a small solar water heater installed at Mt. Wilson in 1915. While such devices were primarily intended to explore the new technology, they also served to publicize the possibility of solar power.

By 1940 Abbot had expanded his inventory to include solar-powered saltwater distilleries, toy stoves, ice makers, and solar broilers. But in the pre-war era of cheap fossil fuels and electricity, Abbot was unable to interest manufacturers in the commerical potential of his dream.

Charles Abbot retired in 1944, confident that his long record of daily solar observations constituted overwhelming proof of the connection between solar radiation and climate and weather suspected by Langley over half a century before. Even so, there had already been intimations that astrophysical studies were changing direction, impelled in part by the stunning advances in physics during the 1920s and 1930s. And when Abbot sought an additional $200,000 for a series of new mountain-top solar stations prior to World War II, Congress ridiculed the

King of the Castle, Secretary Emeritus Abbot—91 years old—peers from his tower office at the Smithsonian's Administration Building in 1963. He claimed the lofty perch in 1928 as the Institution's fifth secretary. Until his death at age 101, Abbot continued to experiment with the sun's energy and to study the relationship between solar radiation and climate.

project. The faith that Representative Cannon had once expressed in the utility of Smithsonian solar research had vanished.

Nor was a new generation of scientists willing to come to Abbot's defense. Astronomers were quick to praise the accuracy of the solar radiation data and the diligence with which it had been collected, but it was obvious that many scholars considered the work to be an anachronistic holdover from an earlier epoch in American science. While his critics admitted that the record of painstakingly collected solar observations had some intrinsic value, they rejected outright any claim that a continued study of solar variation could be used to predict the weather. Abbot found his work compared to a "large and beautiful pearl, admirable, but useless." With Abbot's passage from the scene, the character of the Smithsonian Astrophysical Observatory changed radically as its staff members rejoined the mainstream of astrophysics.

Abbot died in 1973 at the age of 101, having resolutely continued to collect solar measurements until very near the end. His work is now emerging from the scientific limbo where it remained for the middle part of this century. And once again, solar studies have become an important element of the SAO (now the Center for Astrophysics) research program. Abbot's dream of sending sophisticated instruments beyond the Earth's sheltering atmosphere has been realized as sounding rockets, unmanned spacecraft, and U.S. and Soviet astronauts have opened entirely new avenues to the study of the Earth-sun relationship. In cooperation with Harvard astronomers, Smithsonian scientists have established a joint Langley-Abbot Research Program that combines current theory with spacecraft observations of solar phenomena and the unique historical record compiled by Langley and Abbot in a long-term study of solar variation and the sun's impact on the Earth's climate.

On the practical, experimental side,

too, Langley and Abbot left a legacy to a now energy-conscious world. Modern solar power systems, descendants of those small, crude solar-powered "novelties" and "gimcracks" developed by Abbot and others, have assumed an importance today that underlines our awakening sense of the vital kinship between sun and Earth. The Smithsonian emphasis on understanding the way in which the Earth and sun operate as an integrated ecological system led to an early appreciation of the fragile balance that sustains life on our globe.

Langley and Abbot would have been gratified but scarcely surprised by the enormous public response to photographs taken by the Apollo 8 astronauts in December 1968. For the first time we saw our home planet as a delicate blue ball, flecked with white, suspended over a desolate lunar landscape. Suddenly we spoke of "spaceship Earth," wondered at the damage done by her unthinking passengers, and sought a broader vision of ourselves and our planet.

Nearly a century ago Langley had voiced similar concerns in an attempt to explain the deeper motives that guided his solar studies. He spoke of a vision of the Earth and its inhabitants as a unified whole, a vision that is the real legacy of 60 years of Smithsonian astrophysical research. The Earth, he wrote in 1891, is:

The only planet in which man is certainly known to exist, and which ought to have an interest for us . . . for it is our own. We are voyagers on it through space, it has been said, as passengers on a ship, and many of us have never thought of any part of the vessel but the cabin where we are quartered. Some curious passengers (these are the geographers) have visited the steerage, and some (the geologists) have looked under the hatches, and yet it remains true that those in one part of our vessel know little, even now, of their fellow voyagers in another. How much less, then, do most of us know of the ship itself, for we were all born on it, and have never been off it to view it from the outside.

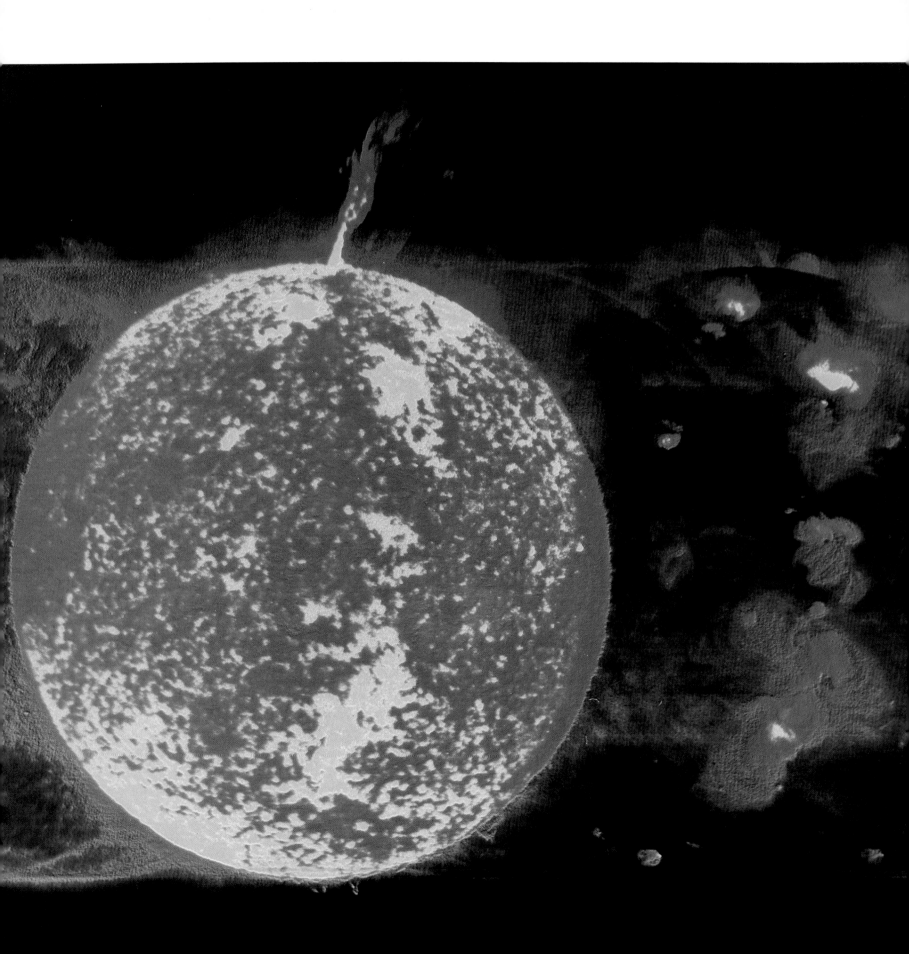

William Livingston

Eyes on the Sun

The bright orange sun dominating this side-by-side image of layers of the solar atmosphere reflects the abundance of helium in the sun's chromosphere. Other "ghost" suns represent scarcer elements in the corona, tens of thousands of kilometers above the chromosphere. The sun's image is turned on its side and the flare at the top is actually near the equator.

Throughout most of human time, the sun has appeared to man as a bright object, usually white, sometimes yellow or reddish, perfectly sharp, perfectly round. While revered as the giver of life, the image of perfection by nearly every culture, the sun's nature remained a mystery to even the greatest of the ancient astronomers.

Only with the invention of the telescope did the sun begin to yield its secrets. Galileo's primitive instrument revealed blemishes—sunspots—and shattered the myth of the sun's perfection. From the transit time of these sunspots he deduced the sun's rotation period, some 26 days. Sir William Herschel, in the 18th century, discovered that the sun emitted radiation invisible to the naked eye. In the 19th century the pace of solar "exploration" quickened, and hosts of instruments were devised to detect, measure, and record the sun's rays. By the eighth decade of this century, solar physicists, using marvelously subtle and sensitive instruments, have been able to answer most of the questions about the sun's nature and source of energy.

This, then, is the story of the instruments, the tools that broaden and sharpen our vision of our own bright star. First, we should explain how we "see" the sun, whether fleetingly with the naked eye, or at greater leisure with instruments. The sun emits energy, and just as sound energy can vary in pitch, or wavelength, solar energy is emitted at various wavelengths. Some wavelengths are visible and we see them as light. The nature of those wavelengths, or frequencies, of light (and of all the other wavelengths of solar energy, too) has long perplexed experimenters, because light seems sometimes to behave like particles and other times to act like waves. Today, we think of light as having some of both properties. Little trains of waves, each finite in length, act like particles—and each wavetrain is called a photon. Astronomers need photons at all their various wavelengths in order to obtain accurate measurements of solar phenomena. Simply put, the more photons that can be collected by instruments, the greater the accuracy of the measurements.

All of the wavelengths of solar energy together comprise the solar spectrum (see pages 50–51) of which visible light is just a small segment. The whole solar spectrum extends from the very short wavelengths of gamma rays and x rays through the ultraviolet region to visible light, and from there into the longer wavelengths of infrared and radio. In designing and using his instruments to detect and measure these various wavelengths, the solar astronomer must take into account several limitations. One such problem is heat. Even a simple magnifying glass can concentrate solar energy enough to burn a hole in a piece of paper. Solar energy falling on a collecting instrument such as a telescope produces a hot image at the focus. The solution to this problem is to avoid concentrating the solar image any more than necessary.

Another problem concerns our atmosphere. For us and all other living things the atmosphere is a very good thing because it absorbs much harmful solar radiation, such as gamma rays, x rays, and the ultraviolet. But the atmosphere's protective quality also prevents the Earth-based solar astronomer from collecting and studying those energetic wavelengths. In addition, atmospheric turbulence distorts light, making stars appear to twinkle and creating the mirages familiar to desert travelers. Such turbulence reduces the clarity of the solar image projected by a telescope, obscuring many details of the sun's features. Astronomers refer to the clarity with which celestial bodies can be observed as "seeing," and the less turbulence and distortion there are, the better the seeing. Placing a telescope into orbit in the natural vacuum of space provides the ultimate answer to the atmospheric problem, although the thermal extremes of space present new difficulties. But solutions are being found, and a solar telescope with a collecting mirror 30 inches in diameter is planned for the Space Shuttle era.

The telescope, however complicated or technically advanced, is just a collector. At the heart of every solar instrument is a sensor or detector which counts the photons and a computer which records the count. Until the 1970's

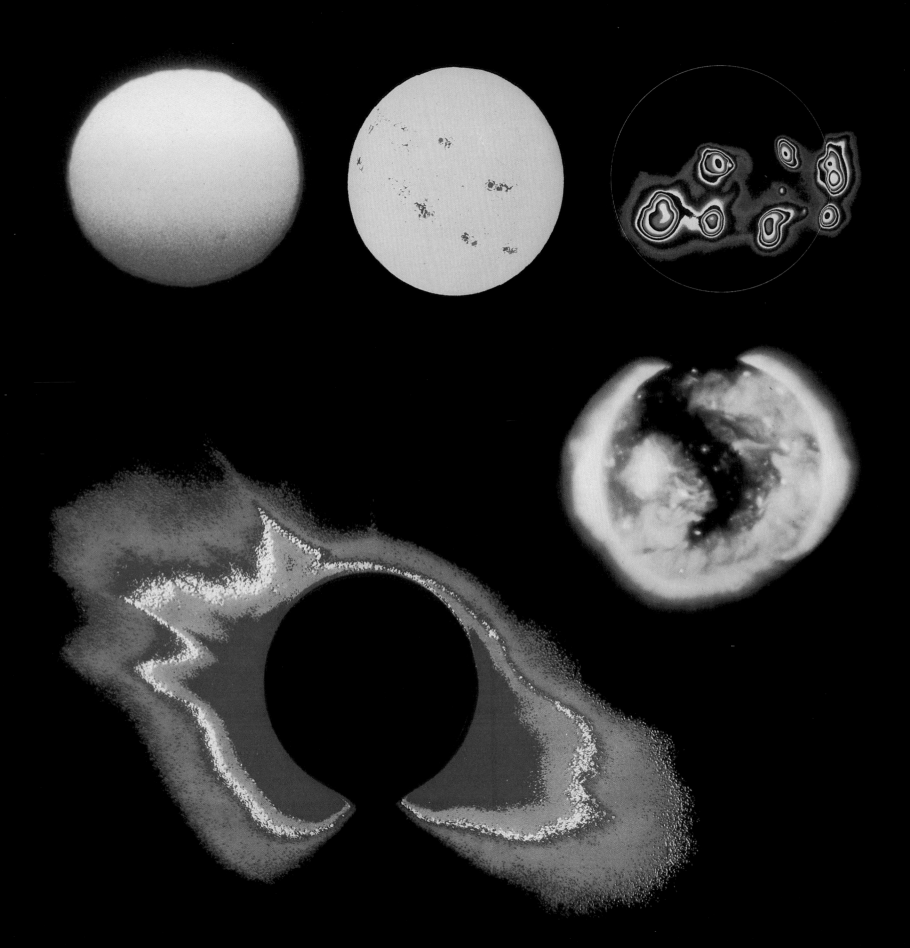

Scientists employ a variety of instruments to peer at—and into—the sun. Telescopes coupled to sensors and computers all but peel away the sun's layers to measure temperature, pressure, and magnetic fields. Images of the sun are studied for clues to its structure, mechanisms, and variability. Most of the images at left have been enhanced and colored by computer to bring out subtle details. Top row, left to right: the white light sun we see on a bright day; a magnetogram, or magnetic map of active regions on the sun, showing intense magnetic fields associated with sunspots; a Skylab x-ray telescope image of violent activity in regions of high temperature and magnetic flux. (Poor penetrators of Earth's atmosphere, solar x rays can be detected only in space.); magnetograms of the sun at a time of maximum activity (left) and minimum activity (right). Middle row: another Skylab x-ray image—bright spots are areas of high temperature; a white light photograph of the sun's corona, or outer atmosphere, visible from Earth only during eclipses. Bottom row: a short burst of furious activity in the corona was revealed in an artificial eclipse created by a Skylab instrument called a coronagraph; the high chromosphere layer of the sun's atmosphere seen in ultraviolet wavelengths from Skylab; another image of the sun in the ultraviolet with a hole in the corona, dubbed the Boot of Italy by Skylab scientists. This feature also appears in the image at the beginning of the middle row.

Viewing the sun may cause permanent damage to the eye, and that damage can be instant and massive if any optical device is used. For safety, consult an ophthalmologist.

astronomers had to contend with such terribly inefficient devices as photographic film which misses 99 out of every 100 incident photons, or relatively effective but bulky systems such as television cameras which can weigh several hundred pounds. Now all that has changed. Responding to virtually every photon, tiny solid-state arrays made of silicon are revolutionizing observational astronomy. Known as diode arrays, these image transducers or converters are a product of the same technology which can place an entire computer on a single chip, or crystal, of silicon.

A telescope, diode array, and computer system is analagous to the human eye and brain system. The lens of the eye focuses light on the

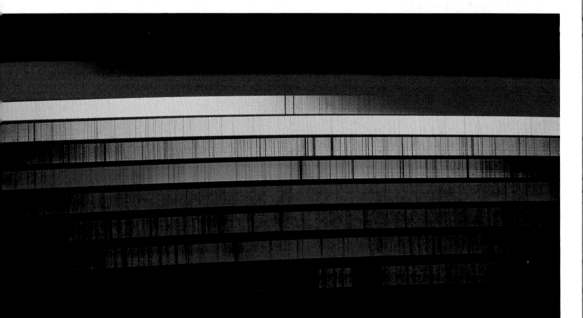

Below, the visible or white light portion of the solar spectrum has been split into all of the colors in this spectrogram from Sacramento Peak Observatory, New Mexico. Dark lines on the colors mark ionized atoms of the sun's elements. Right, the author obtained this spectro-

rods and cones of the retina. Each rod or cone senses the light falling on it and sends a message to the brain via the optic nerve. The brain in turn organizes all of the messages and produces an image. Similarly, the telescope focuses light on the diode array's pixels, which send their photon counts to the computer. The computer processes the signals and organizes them into a numerical readout of what the telescope "sees." By using many such diode arrays, observations requiring accuracies that previously could only be dreamed of are now feasible.

There's more to the sun, however, than photons, and for other observations instruments in addition to telescopes are needed. Spawned by an unseen generator deep in the sun's interior,

gram of the sun's "surface" using the McMath Telescope at Kitt Peak National Observatory, Arizona. The vertical dark line represents a small sunspot, while the horizontal green line is hydrogen alpha. Most other horizontal lines are metals such as iron, nickel, and vanadium. Far right, a coelostat, a device with a clockwork-driven mirror that tracks the sun across the sky and reflects its light into a fixed instrument.

magnetic fields continually break through the solar surface. When concentrated, the fields coincide with sunspots, but whether concentrated or dispersed, magnetic fields appear to be the basic progenitors of solar activity. Mapping and tracing the evolution of surface magnetism is presently a preoccupation of many solar observers. Magnetism, they hope, will prove a key to understanding the sun itself. Mapping magnetism requires a large vacuum telescope coupled to an instrument called a magnetograph, which contains a silicon diode array detector. Atoms abundant in the sun's atmosphere, such as hydrogen, helium, and iron, absorb light and reradiate it at discrete wavelengths depending on the element. Broken down by appropriate instruments, these appear as "absorption lines," and they form the solar spectrum. The magnetograph measures shifts in the absorption lines induced by intense magnetic fields, providing data from which velocity and magnetism maps of the solar disk can be prepared.

Although the sun is a stable object, some scientists suspect it may be weakly pulsating or perhaps, more correctly, ringing like a bell. If the sun's "surface" were regularly expanding and contracting, such a heaving would change the velocity of atoms in the sun's atmosphere in relation to an Earth-bound viewer, and should produce shifts in the signals detectable with a magnetograph. A group in the Crimea, using a magnetograph, appears to have found a 160-minute beat with an amplitude of a few meters per second. Recently, this observation was confirmed by French and U.S. astronomers at the South Pole. University of Arizona physicists claim they see the edge of the sun's disk vibrating with periods of 20–40 minutes. But the prize, hands down, must go to English researchers from the University of Birmingham. They find five-minute oscillations across the whole disk with an amplitude of about 20 centimeters per second! We will return to the significance of this result later.

The eruption of magnetic fields through the solar surface has an 11-year periodicity called the solar cycle. Solar minimum last occurred around 1976. Magnetic activity rather quickly rose to a maximum in 1979–1980, then slowly began to trail off. Since all manifestations of activity, namely sunspots, flares, prominences, and ultraviolet enhancement, are magnetic in origin, and since ultraviolet and other rays—gamma and x rays in particular—are best

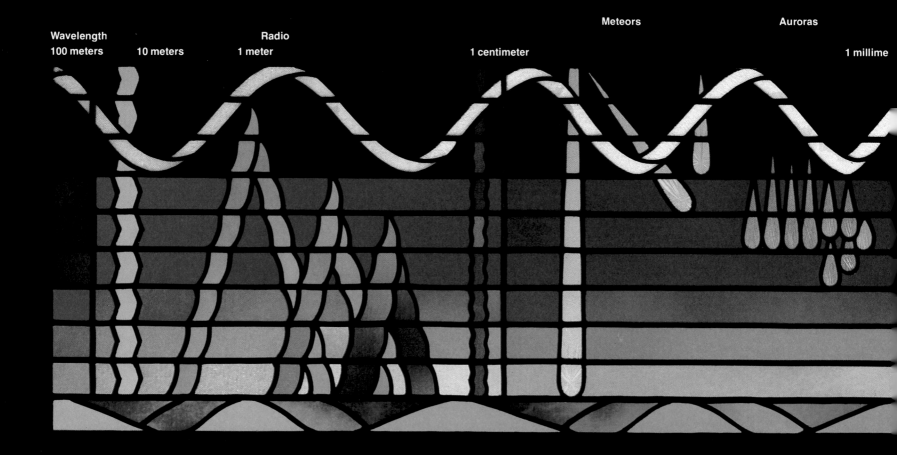

**Wavelength
100 meters** **10 meters** **Radio
1 meter** **1 centimeter** **Meteors** **Auroras** **1 millime**

Spectrum of Solar Radiation

The electromagnetic spectrum of the sun consists of the entire range of energy emitted by the sun at all wavelengths. Electromagnetic radiation can be thought of as traveling in waves, and the lengths of the waves vary considerably, ranging from over 1000 meters long to about 0.0001 of an angstrom unit, (Å), or one ten-trillionth of a meter. From long wavelengths to short, these include radio, infrared, visible light, ultraviolet, x ray, or gamma ray. Radio waves, shown above at the left, penetrate the Earth's atmosphere and reach the surface, where they can be detected as solar "static" by radio tele-

Infrared	Visible Light	Ultraviolet	X ray	Gamma Ray	Cosmic Ray
100 micrometers	1 micrometer	100 angstroms 10 angstroms		0.1 angstrom	0.001 angstrom 0.0001 angstrom

scopes. Meteors flashing through the ionospheric layer high in the atmosphere create radio "noise" around these wavelengths as do auroras in the ionosphere—these caused by solar activity. Next is the infrared region. While infrared radiation cannot be seen, it can be felt as heat on the skin. Merging with infrared is the visible portion of the spec-

trum, the light by which we see, carrier of much of the solar energy reaching the Earth's surface. Beyond this rainbow of visible light lies the region of the ultraviolet. Ultraviolet rays, responsible for sunburns, would be lethal if they were not largely screened by our atmosphere. Even more dangerous x rays and gamma rays are completely

absorbed by the Earth's atmosphere and can only be studied from vehicles in space. At the short end of the spectrum are the cosmic rays, ultrashort wavelength radiation from the sun as well as other stars. This depiction of the electromagnetic spectrum of the sun was created in stained glass by Jackie Leatherbury Douglass.

detected above the atmosphere, this was an ideal time to use a satellite. Hence the Solar Maximum Mission satellite (SMM).

International in concept, involving some 50–70 solar physicists from Europe, Japan, Australia, and the United States, SMM was launched by NASA in February, 1980, and directed from Goddard Space Flight Center near Washington, D.C.

One group operated an ultraviolet magnetograph to map magnetic fields in the hot chromosphere where flares probably originate. Solar flares which last only minutes or even seconds emit copious quantities of gamma rays and x rays. During major flares, ejected particles are accelerated to very high velocities and energy levels. If the flare site is just to the west of the sun's center, the particle trajectory will include the Earth and we will then experience a geomagnetic storm. SMM instruments called spectrographs, sensitive to these radiations, were used to detect flares with very accurate time resolution. The flare acceleration mechanism is not understood, and the hope is that these spectrograph timings will provide clues.

Another group is concerned with gamma rays. These are very short wavelength x rays which pass right through ordinary mirrors, so special telescopes must be used to collect them. Because x rays have wavelengths less than a hundredth that of visible light, the telescopes can be small in aperture yet attain a high resolution.

Another SMM package is a coronagraph. The intensity of the corona surrounding the sun is about a millionth that of the solar surface. By covering the sun with a disk held several feet in front of the telescope and by taking special care to have dust-free lenses to reduce scattered light, an artificial solar eclipse can be produced, and the tenuous corona appears clearly against the blackness of space. A similar instrument aboard Skylab in 1973 saw the sun eject a number of strange expanding bubbles dubbed "coronal transients." If such bubbles can be tied to chromospheric phenomena, then the emergence of the latter can be established as a precursor and thus a predictor of coronal transients.

These were only a few of the instruments carried on SMM. Others included a radiometer which monitored the sun's energy output, or the solar constant, to better than 0.1 percent accuracy. Smithsonian secretary and solar physicist Charles G. Abbot (see pages 38–43) believed he had observed variations of the solar constant of approximately one percent, appar-

ently connected to solar activity. SMM should clarify any such behavior; to date it has recorded fluctuations at the 0.1 percent level.

While some solar instruments are most effectively employed in space, others must be buried deep in the Earth to do their collecting. According to theory, four percent of the sun's energy output is in the form of neutrinos. These elusive "little neutral ones," as physicist Enrico Fermi called them, have little or no mass or charge and pass through matter practically unimpeded. Created as an essential by-product in the fusion furnace of the sun's interior, only one in 100 billion is captured during its flight from the core to the solar surface. At the distance of the Earth from the sun, 10^{13} neutrinos pass through your body every second. Although they are hard to stop, their very numbers make possible the capture of a few. A neutrino "telescope," consisting of a 100,000-gallon tank of tetrachloroethylene—a cleaning fluid—buried a mile deep within the Homestake gold mine in South Dakota captures two neutrinos every six days. But—assuming our recipes for atomic reactions are correct—acceptable theoretical models of the sun's interior predict the capture of five neutrinos every six days, instead of two,

according to Raymond Davis, Jr., who designed the experiments. Recalculations and reassessments of the atomic processes have not closed the gap between theory and observation.

Help now appears from an unexpected quarter. If the 20-centimeter-per-second pulsation detected by the previously mentioned group in Birmingham, England, proves valid, it could mean our accepted model of the solar interior is in error in that we assume the wrong mixture of chemical elements within the sun. After a suitable correction, the standard model predicts, *voilà*, two neutrino captures in six days!

What else might we expect to see happening in solar astronomy? Again, observation from space offers so many advantages along with some disadvantages.

Today, if I have an inspired idea, I can be on the telescope tomorrow trying it out. On the other hand, an experiment in space involves a five-year lead time and elaborate team efforts. Potentially, Space Shuttle facilities solve both problems. First, using the Shuttle, the large and bulky high resolution telescopes can be assembled in space and then, most important, fine-tuned for optimum performance. Second, the equipment will be regularly serviced, perhaps even monthly, so that fast response, individually inspired astronomy—in space—will become practical. Perhaps with these tools the sun will at last be forced to reveal its innermost secrets. At the least, we can be certain that a long line of solar astronomers will be waiting at the Space Shuttle ticket office.

The McMath solar telescope, above, stands over 30 meters (100 feet) high near the snowy summit of Kitt Peak. The diagram at right traces the path a beam of light must take through the McMath to reach its destination. A heliostat, or tracking mirror, at the top of the shaft reflects light down to the 1.5 meter (60 inch) primary mirror. Concentrated, the beam bounces back to a secondary mirror that directs it down into a receiving instrument. The beam may also end up at the solar table, left, to the delight of a group of Navajo students. An image of the sun 77 centimeters (30 inches) across and four times the intensity of normal sunlight, can be projected.

McMath Solar Telescope

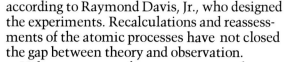
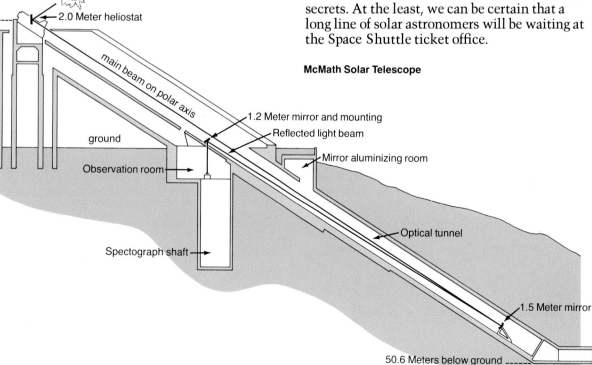

Sun

2.0 Meter heliostat

main beam on polar axis

ground

1.2 Meter mirror and mounting

Reflected light beam

Observation room

Mirror aluminizing room

Spectograph shaft

Optical tunnel

1.5 Meter mirror

50.6 Meters below ground

High in the Pyrenees, domes of France's Pic du Midi Observatory shake free of massive snowdrifts. Astronomical seeing is splendid here, and the Pic's astronomers make the most of it both day and night, in particular obtaining some of the highest-resolution pictures in existence of the solar face. Two such astoundingly clear photographs appear in the story beginning on page 84. The main point of this section, however, is that all astronomy—day or night—deals with suns and sun systems: for all the stars are suns, kin to our own daystar.

Our astronomer authors come down from their mountains and their telescopes to cast poetic eyes "from heaven to earth, from earth to heaven" to reveal the unfolding pattern of celes-

A Place in the Cosmos

tial creation. They build cosmic context and perspective and definitely fix our planet's location within the grand scheme of it. The awesome notion of how all matter, space, and energy fit together suggests a chronology that can be examined not only from its inception but to this very day. We can even predict a future course of events for our home-star and its galactic companions.

George Field of the Harvard-Smithsonian Center for Astrophysics peers down the long, long passages of time to see how galaxies and stars are born and includes a brilliant discussion of the Big Bang. Even Einstein's equations break down when confronted with the beginning of the universe.

Astronomer Kenneth L. Franklin and noted illustrator Helmut K. Wimmer recreate the inception, early development, and eventual ignition of our sun and birth of our planets. Franklin's scenario reveals that the great Jovian planets gained much of their present form before the sun began to shine, with the small inner planets losing material during a gravity storm that heralded the new sun's first gleaming. With the solar system's overture thus played, David Morrison explores the differing paths of planetary evolution. With word and picture he celebrates the remarkable era of the 1960s and 1970s during which astronomy changed from a spectator sport to a field event, with man's first steps on the moon and the mechanical marathons of unmanned spacecraft sent out to reach the planets.

George Field

This Special Galaxy

Seeing should be believing: Andromeda, our nearest galactic neighbor, appears to float in a sea of stars. Only it doesn't. Most of those stars are actually part of our own galaxy—the Milky Way. They form a troublesome screen through which the earthbound observer must peer. No stars pave the empty reaches between Andromeda and the Milky Way. Alone, the "island universe" swirls in its glory, 1,800,000 light years away.

The blessed singleness of the Andromeda galaxy is important in our understanding of how all galaxies, including our special one, came to be. To get a feel for the way galaxies are arranged in space, let's just for a moment compare stars to roses: then galaxies would be widely scattered rosebushes packed with buds and flowers.

In spite of fascinating varieties and hybrids, all of Creation's 10 billion galaxies would seem to be cloned from the same parent stock, and all at the same time. Most astronomers believe that it happened 10 to 20 billion years ago. For simplicity, we adopt a specific value, 20 billion years. And the event, the Big Bang, required only 180 seconds of creative chaos to send everything hurtling outward from nowhere on a journey that has led to our sun, to today, and presumably will continue long after we and our sun are gone.

It is the purpose of this article to fix the sun's place within the scheme of cosmic creation. And to do so without some preliminary qualification is both misleading and deceptive.

In the first place, the Big Bang is an intellectual abstraction, a model, but even if the origin of the universe didn't happen just as described, this metaphorical device is nevertheless most useful. It is a kind of vehicle which enables the brain to travel the long, long corridors of univer-

Like fuzzy moons, two dwarf elliptical galaxies hover near the starry pinwheel of Andromeda, a giant resembling our own Milky Way. Though varying in size and detail, such galactic bodies hold almost all of the universe's matter. Much of the energy, however, lingers in deep space as heat left over from the Big Bang.

57

sal time and space. In this it shares the qualities of a work of art, a symphony, for instance.

Prelude

The ancient science of astronomy deals with a universe unimaginably more ancient. Like the explorers of the Renaissance period, astronomers have ventured on uncharted journeys into space, hoping to understand something of our silent universe. They have now charted the celestial continents; they have been followed by astrophysicists, carrying tools of physics fashioned in earthbound laboratories, seeking to discover how the great structures in deep space—stars and galaxies—are built, and even how they came to be.

With this in mind, let us make an enlightening journey. We move beyond the Big Bang to a day not long ago when the idea did not yet exist in scientific minds. And let us keep in mind that without some concept such as the Big Bang, we could not visualize the relationships among stars and galaxies.

Unlike the physicist, the astrophysicist is unable to construct a laboratory large enough to carry out the experiments he would like to do—bore deep into a star, pluck a galaxy, try another universe. But he can use numbers and computers to fashion models of stars, galaxies, and even the universe. Stimulated by the data streaming in from our optical, radio, and x-ray telescopes, he attempts to reconstruct the history of the universe and of all the various structures within it.

Sir Isaac Newton showed the way by using a prism to break up sunlight into its component colors. Astronomers then turned prisms to the stars to determine their surface temperatures and their chemical composition from the spectra. And there are tremendous numbers of stars. We can see about 5000 with the unaided eye, and millions of them with the largest telescopes.

Radio astronomers with their big electronic dishes can literally penetrate the dust clouds of the Milky Way and statistically "count" our 200 billion stars. And while dust prevents us from seeing our own galaxy in all its glory, we can look at Andromeda, our sister galaxy. As we do, we can appreciate how an intelligent being living far out in Andromeda's disk would see a band of light he might call *his* Milky Way.

Beyond nearby Andromeda, astronomers discovered other galaxies in great numbers and forms. Most of them are spirals like our own,

and also rounded giant ellipticals. Astronomers estimate that the most distant galaxies so far observed are several billion light years from the Earth.

All this fits in with a great discovery that Edwin Hubble made in 1929. He determined the speeds of the galaxies by observing the degree to which their starlight shifts in wavelength—the famous red shift. He found that almost all galaxies are receding from us at significant fractions of the speed of light. Numerically, this factor is between 16 and 32 kilometers per second for each one million light years' distance between the Milky Way and any other particular galaxy. Such figures imply that billions of years ago the galaxies must have been much closer, perhaps even touching.

The discovery of Hubble's expanding bubble was foreshadowed by a little-known Russian mathematician, Aleksandr Friedmann, in 1922.

Crab Nebula's bright claws represent the expanding shell of a supernova—the explosion of a large star. Such thermonuclear fury helped transmute lightweight hydrogen and helium into the heavy elements upon which life depends on our planet (see pages 64 and 81). Our own sun may end with the formation of a planetary nebula, a gentler event in which a dying star envelopes itself in a gaseous shroud rich in carbon and oxygen.

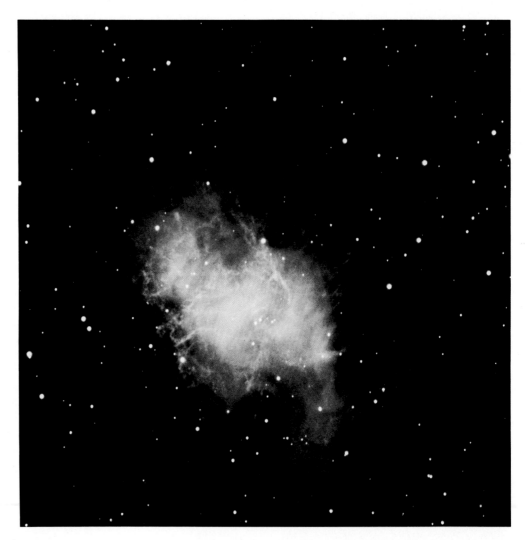

Massive Triffid Nebula glows darkly through interstellar gas clouds—wombs for stars. These are laden with cosmic dust grains, and cold protostars gain bulk here. Later they flare into life from nuclear reactions like those that lighted our sun five billion years ago. Heating occurs first through gravitational compression of gas and dust, with nuclear fusion taking over after temperatures reach 10 million degrees F.

Working with Einstein's general theory of relativity (put forward in 1915 as a revision of Newton's laws of motion and gravitation) Friedmann had asked himself whether the new hypothesis suggested any simple model for the universe as a whole. He found that it did.

An important and disturbing feature of the Friedmann model is that the universe's expansion started with all the matter in one single point. At the instant of the beginning it became a volume of innumerable points, expanding with great speed to make the space we perceive today as the universe. Such an explosive emergence might deserve a title like Big Bang, but giving it a name does not resolve the enigma of what caused this genesis in the first place. Science is silent. Einstein's theory—which seems to provide an adequate account of the universe once it has started to expand—breaks down when its equations are applied to the initial instant of time. We may, however, place ourselves within the primal enigma.

Initial Theme

In a single millionth of a second, our universe—and time itself—begins. All the energy around us has sprung out of a region so dense and so hot that—as we know it—time had no meaning. At this moment the temperature of the gas around us is 10 trillion degrees, but it drops perceptibly with each passing microsecond. The entire universe is only a fraction of a kilometer across, but it expands at huge speeds —matter quite close to us being propelled at almost the speed of light.

At the tremendous temperatures involved, the matter in the universe consists of elementary particles—photons, electrons, neutrinos, mesons, and quarks. Some of these, like photons and electrons, are familiar. All of them are

accompanied by their antiparticles, shadowy mirror images which they annihilate on contact. No sooner does a particle-antiparticle pair annihilate, however, than another is created out of the tremendous amounts of raw energy available in the early universe.

The time is now 10 microseconds, and the temperature has dropped to three trillion degrees. Under these conditions fundamental particles called quarks and antiquarks have begun to stick together in various combinations to form protons, neutrons, and their antiparticles. With the further drop to one trillion degrees at a time of 100 microseconds, all the quarks have been used up in making protons and neutrons, and the antiquarks in making antiprotons and antineutrons. The proton-antiproton pairs and neutron-antineutron pairs are constantly annihilating, but now the universe is so cold that the energy for making new pairs to replace the old ones is hard to find.

The result is a massive execution of just the type of particles—protons and neutrons—we were counting upon to make the sun and, for that matter, the stars and galaxies throughout the universe. But wait—a few protons and neutrons are surviving! For some still inexplicable reason there was a slight excess—about one part in a billion—of matter over antimatter right at the start. Since there is not quite enough antimatter to annihilate all the matter, it looks as if the universe will retain a little stuff after all.

Now the universe is a hundredth of a second old and there will be a tomorrow. By the time the universe reaches the advanced age of one second, the temperature has dropped to a comparatively frigid 10 billion degrees. The ratio of protons to neutrons is 76 percent to 24 percent. Three seconds later, the temperature is five billion degrees. There is no longer enough energy to create electrons and antielectrons. So the leftover opposing pairs annihilate just as the proton-antiproton pairs did before them. Fortunately, again, electrons barely outnumber antielectrons. It is just enough to guarantee one electron for each proton—and that is exactly what is needed to produce neutrally charged atoms as the universe cools off rapidly.

After only three frenetic minutes, the Big Bang is virtually over! The emerging universe contains approximately 73 percent hydrogen and 27 percent helium, plus minute fractions of deuterium and light helium. It looks as if this composition will hold from here on out.

The Bang, as we observe it, proves to be Big only in terms of its unutterable output of energy and matter. The explosion itself is physically small enough to fit inside the orbit of Mercury, the innermost planet. Theory suggests that once the primal nebula was born, virtually all else became a matter of diffusion and the independent evolution of the galaxies.

The nascent universe around us glows brightly, but there is a kind of cosmic fog which prevents us from seeing very far. Electrons created in the Big Bang scatter the light just as tiny water droplets scatter the sunlight through an early morning mist. So the glow from distant parts never reaches us: we can see only the nearby gas. This foggy situation persists through 100,000 years while the universe goes on expanding and cooling.

As the millennia pass, temperatures drop, and for the first time the hydrogen and helium atoms—which form when nuclei attract electrons to them—can survive the universe's intense radiation. For the first time, atoms begin to outnumber bare, battered atomic nuclei in the universe.

At last the fog clears. Because the electrons now combine into atoms, nothing remains to

Swirl of stars and cosmic clouds resembles melting cotton candy or a stir of seed-filled berry jam. This artist's conception presents two views of our own galactic disk: one full face, the other in thin profile. Cutaway of the profile emphasizes the structure of galactic arms. Our sun, a yellow star occurring on one arm, would be all but invisible at this scale if not highlighted. The Milky Way's bulbous center holds cool and unexceptional stars. Observations of the very heart, however, suggest the presence of a Black Hole, a powerful gravity sink (see page 37 for an illustration). Far right: sky full of rainbows provides scientists with clues to the condition and composition of stars. A prism attached to a telescope projects each star's characteristic spectrum of colors and other features that indicate temperature and chemical makeup.

scatter the radiation, and for the first time we can see deep into the universe. More important, matter can now finally begin to accumulate. Let us watch closely. Bound by their own gravitation, masses of this foundation stuff tend to dissipate less rapidly than the surrounding gas. Shapes are becoming ever more prominent. The first suns are waiting to be born.

The smooth distribution of matter from the Big Bang no longer exists. Distinct entities arise, each cloud containing the mass equivalent of roughly 100 billion suns. Internal gravitation has now overcome the original tendency

of their gases to expand along with the rest of the universe and, in one spectacular case, a collapsing cloud suddenly glows. Stars will be born out of the hydrogen and helium left by the Big Bang. Eventually, as the large cloud contracts, smaller clouds within it also become unstable and collapse upon themselves.

As we watch, that vast clump turns completely into stars. We are witnessing in a mere billion years the birth of an elliptical galaxy, a rounded shape filled with stars that travel outward from a center and then work their way back inside.

Though our model does not yet deal at length with the issue, it seems clear that none of the first stars should be burning today. For stars—as with biological creatures—go through life cycles. And while the general appearance of an individual galaxy seems to endure for a long time, the particular stars within that—or any—galaxy have likely changed over and over again.

Burning hot and bright, the heaviest stars rather rapidly run out of the hydrogen fuel for their nuclear reactions. As we watch, the most massive giant stars in the elliptical galaxy become unstable, swell enormously, and explode as supernovas. Temperature and pressure act to forge for the first time the atomic nuclei of such heavy elements as silicon, magnesium, and iron.

Slow-burning stars of lighter weight follow a different life history. They, too, swell to become giants but, instead of throwing off matter in a supernova explosion, they puff their gas out gently, forming misty envelopes known as planetary nebulas.

There ensues a dramatic interplay between the outrushing supernova shells and the more sedate masses of gas and dust. These include the significant carbonaceous material formed in planetary nebulas. Supernova shells are traveling so fast that they simply drive any such clouds in their way right out of the rounded elliptical galaxies. Little matter accumulates here.

Counterpoint

Once again we look upon our condensing universe, but this time to witness the formation of a galaxy like our own. Rather than contracting in the spherical fashion of ellipticals, spiral galaxies quickly become flattened due to the rapid spin imparted to them at birth. Apparently the primordial gas was turning faster where spiral galaxies formed than where elliptical ones originated. As a result, the stars huddle closer together in the flat, wheeling spirals than in bulging ellipticals.

When the massive stars explode as supernovas inside spiral galaxies, the rapidly expanding shells of gas find themselves trapped by dust clouds and planetary nebulas—with a total weight that is too great to be pushed out of the galaxy. Thus in a spiral galaxy the material cast out by dying stars is retained to form a permanent interstellar medium, one rich in carbon and oxygen, and one ultimately capable of nurturing life. Such were our origins.

Internal Harmony

Galaxies have already drifted far apart, each undergoing evolutionary development. Those like our own provide a beautiful sight, their luminous spiral arms delineated by newborn stars . . . like beads on a string.

The pinwheel pattern of our spiral galaxy is believed to be the result of vast rotating sound waves. Pressure from the moving waves compresses streams of interstellar gas into clumps, the gas gathering into giant clouds of a million solar masses or more. These clouds contain a seminal blend of hydrogen molecules, dust, and helium atoms.

Affected by their own gravity, the giant molecular clouds begin to collapse as the galactic spiral wave goes by. The result is a burst of newborn stars at the leading edge of the spiral wave, followed by the rest of the oncoming dark molecular cloud.

The sun's birth itself proves unspectacular, for our star emerges in an association of several thousand other stars. They form when a typical molecular cloud collapses at a distance of about 30,000 light years from the center of the galaxy. As the cloud contracts, it breaks up into star-sized blobs of gas, each destined finally to become a star by collapsing on itself.

One clump will become our sun. Endowed at birth with a small amount of spin, it whirls ever faster as it collapses. Like a lump of dough spun into the air by a deft cook, it flattens out.

Gas gathers at the center of the flat protoplanetary system, still lit only by light from distant stars. With each turn of the disk, the weight of its center increases, compressing and heating its interior; eventually the hub begins to shine, and a star—the sun—is born as nuclear reactions commence. Thus, a clump of gas which had emerged from the primeval galaxy contracts and ultimately becomes our sun.

The story is told, but only in broad outline. Details of our sun's inception and birth appear on the following pages. Still later, another author ventures a scenario of the sun's long career and demise. Except for hints, the fate of the universe is unforetold. But if galactic history repeats itself, the ashes of our sun will be incorporated in yet other suns, and perhaps in other creatures.

In cosmic terms, there definitely seems to be heaven enough, and time, for this world to pass away and for others to come in their orderly succession, but toward what end our theories can scarcely fathom.

Light pouring through a cathedral rose window brings to mind certain basic features of the latest cosmological theory, and especially the Big Bang concept. Of course there is no direct connection, only an apparent one in the eye of the beholder. The center of this rose window at Saint Bonaventure Cathedral in Lyons, France, seems to rotate—an optical illusion. The movement suggests to some observers the annihilation of matter by antimatter in the opening movements of the Bang. As if spinning out from the center, rounded and wavelike forms remind us of the two basic galactic shapes—rounded, and spirals that rotate. This particular window was made during the 15th and 16th centuries.

Supernova to Star

Kenneth L. Franklin

Illustrations by
Helmut Wimmer

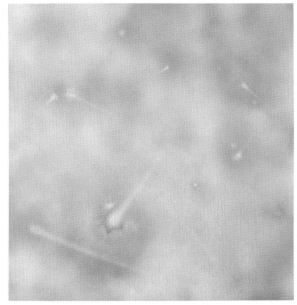

In the previous article we traveled through the galaxies in search of solar origins. Now we settle in close to home and learn in detail of the sun's direct ancestors. We start with a huge cloud of cold gas and dust. Belief is growing today that a supernova occurred near the cloud and that some of its matter penetrated the cloud, infusing it with certain elements produced in the supernova event—its total history revealed in a figure on pages 66 and 67. We'll go into detail on that productive blast after some of the arcane but necessary background that astronomers often conjure up before they get to the payoff, in this case the payoff being our sun, our planet, and ourselves —all life.

To begin: four major events in the history of the universe were crucial to our existence. First, the initial expansion of everything, commonly called the Big Bang. Without this there would be nothing. Second, the development of the chemical elements of which we are all made. Third, the formation of the sun and its planetary system, which provided a place and the proper conditions for the fourth event, the initiation of life on the Earth which led directly—and only recently—to ourselves. Here, we concentrate on the second and third events, first accounting for the chemicals involved in life.

Until the late 1940s, the favored theory for the origin of the universe involved a tremendous explosion in the distant past. This seemed obvious because today all galaxies we observe beyond ours are receding from us; the farther away they are, the faster they are going. By mathematically reversing time—as in running a motion picture backwards—we can see that everything was in one place at one point in time—The Beginning. The theory said that all chemical elements had to be formed in the first few minutes of this Big Bang, when critical conditions prevailed. Perhaps all of the hydrogen of the universe and much of the helium did form then, but it was difficult to account for the rest of the elements.

About the same time, an alternative theory was announced—the Steady-State model— which maintains that the universe has been in a constant state at all places and at all times. Thus, as the galaxies recede, matter in the form of hydrogen atoms is created between them to keep the universe in a state of constant density. A startling feature of the steady-state theory is that the universe has no beginning and no end. Time is endless. And compared with the Big Bang, there was no initial period in which extreme conditions would have allowed the formation of the elements. This theory caused astronomers to seek other means and they found them inside stars.

The Supernova

In the early 1950s, the first great steps were made in understanding how a star forms, runs through its life, and comes to an end. Now, decades later, we are reasonably comfortable in our knowledge that the formation of the chemi-

cal elements is intimately involved with stellar evolution.

A star is formed from a vast cloud of gas and dust. (In the earliest generation of stars, the material may well have been only hydrogen and helium.) Something initiates the formation: perhaps random internal actions, a passing cloud starting a compression toward a center, even the pressure of light from surrounding stars. However precipitated, the cloud's density passes a critical value and the cloud begins to collapse on itself. The more rapid the collapse, the more massive will be the resulting star. When conditions at the center pass another threshold, nuclear reactions begin to release energy and a star is born.

The star we depict will be much larger and hotter than our own star, the sun. But basically its actions and reactions resemble those of all the stars. The overlying gases adjust themselves to the flow of energy from the center, and the star settles down to a long life as it consumes a vast store of hydrogen fuel at its core (see page 86).

From here on we view the star primarily as a great retort for cooking up dense chemical elements, rather than an atomic engine for producing energy. In time, the hydrogen fuel is used up and the core of the star contains the residue of this "burning"—helium nuclei. As the nuclear fires die down, the core itself begins to cool and shrink, and the process of contracting heats the core, but not enough to stop the collapse.

Fresh hydrogen from the region outside the core begins to enter the core region and becomes fuel to continue the conversion to helium. But it is no longer concentrated in the center; it is now burning as a shell surrounding the original core. This old core continues to contract, while the outer parts of the star heat up and expand from the new generation of energy. The surface area expands faster than the increase in energy flow, so the surface cools off: the star is now a red giant. This will happen to our sun in about five thousand million years.

As its core continues to shrink, its temperature, density, and pressure increase. At yet another threshold, helium begins to burn in what is called the triple-alpha process, producing carbon and releasing more energy. Carbon, combining with more alpha particles, brings about the production of new elements such as oxygen, neon, and magnesium. These elements form successive shells of residue around a progressively shrinking shell of actively burning fuel. As this heat-producing layer consumes more of the core—burning down toward the center of the star—more shells of residue accumulate. Finally, iron is produced along with nickel and similar elements. It is final, because the making of any atomic nuclei heavier than iron will not yield energy—in fact, it requires more energy than it yields.

As the fuel that generates iron is depleted, again the core cools, and again it contracts. This crowds its particles, increasing its density. The iron nuclei are squeezed closer and closer

Four phases of creation:
FIRST—The Big Bang, opposite. Current theory suggests that the primal nebula formed some 15 to 20 billion years ago.
SECOND—Intricate alchemy within the stars themselves transmutes helium and hydrogen into the heavy elements.
THIRD—Stellar nebula, enriched with chemicals from dying stars, condenses into our sun and solar system.
FOURTH—Through the application of light and other forms of energy, life emerges from primitive molecules of organic matter.

together until they cross one more critical threshold: the gravitational attraction between adjacent particles becomes irresistible. The iron nuclei, so painstakingly built up during the life of the star, are crushed like apples in a cider press. Neutrons and protons and other nuclear particles are freed; protons and electrons join to become more neutrons. The core collapses in an instant, allowing the overlying material to fall toward the center in a rush. Neutrons fly out to meet the incoming matter and generate new chemical elements on the way. The sudden halt of the collapsing outer layers against the now tiny core heats the matter in a flash, resulting in a tremendous explosion. The star is blown apart: this is a supernova.

Only a core remains behind, consisting of neutrons in a mass a few miles across, spinning several times a second. The density of this neutron core is about a billion tons per teaspoon. The first neutron star to be discovered was that in the Crab Nebula, the remnants of the supernova of 1054. Near the surface of the neutron star, radio astronomers found a bright source of pulsing radio waves, to which they gave the name pulsar.

The outward spreading gases of the supernova explosion mix with interstellar matter, contaminating it with impurities—the newly formed elements of which we are all made.

The Solar Genesis

The existence of the solar system is a fact. And given the changeable nature of things in the universe, there must have been a time when the solar system was not. The general form of the solar system is agreed on today, but there is little agreement on how it got this way. Thus the following scenario may draw as much criticism as approbation.

We pick up the story as the supernova's speeding blast wave heads toward our maternal cloud of cold gas. One effect it may have upon encountering our cloud is to trigger, or to increase, the tendency of the cloud to collapse toward its center. So now we have one cloud penetrated and enriched with heavyweight minerals from a supernova. Some such clouds exhibit low internal turbulence and thus remain in one piece; and this is the kind of cloud that will commit most of its material to forming a star.

Random movements of interior blobs will eventually be cancelled out by encounters and collisions. This process results in a disk of par-

Seven stages mark the life of a star—now perished. With its death our sun was conceived.
Stage 1—A cloud of dust and gas collapses and heats through compression.
Stage 2—By nuclear fusion the blue-hot star produces heat and transmutes light elements into heavy ones.
Stage 3—With hydrogen fuel depleted, helium burns and the star swells into a red giant with production of chemical elements. Refer to legend and colored arcs in the diagram for identification, right and opposite.
Stage 4—Continued buildup of heavy atoms.
Stage 5—With the output of iron, the star's productive existence draws to a close.
Stage 6—Supernova blast hurls material into space.
Stage 7—Only a dense core remains, a neutron star.

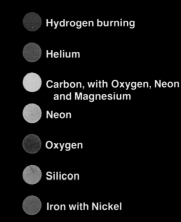

- Hydrogen burning
- Helium
- Carbon, with Oxygen, Neon and Magnesium
- Neon
- Oxygen
- Silicon
- Iron with Nickel

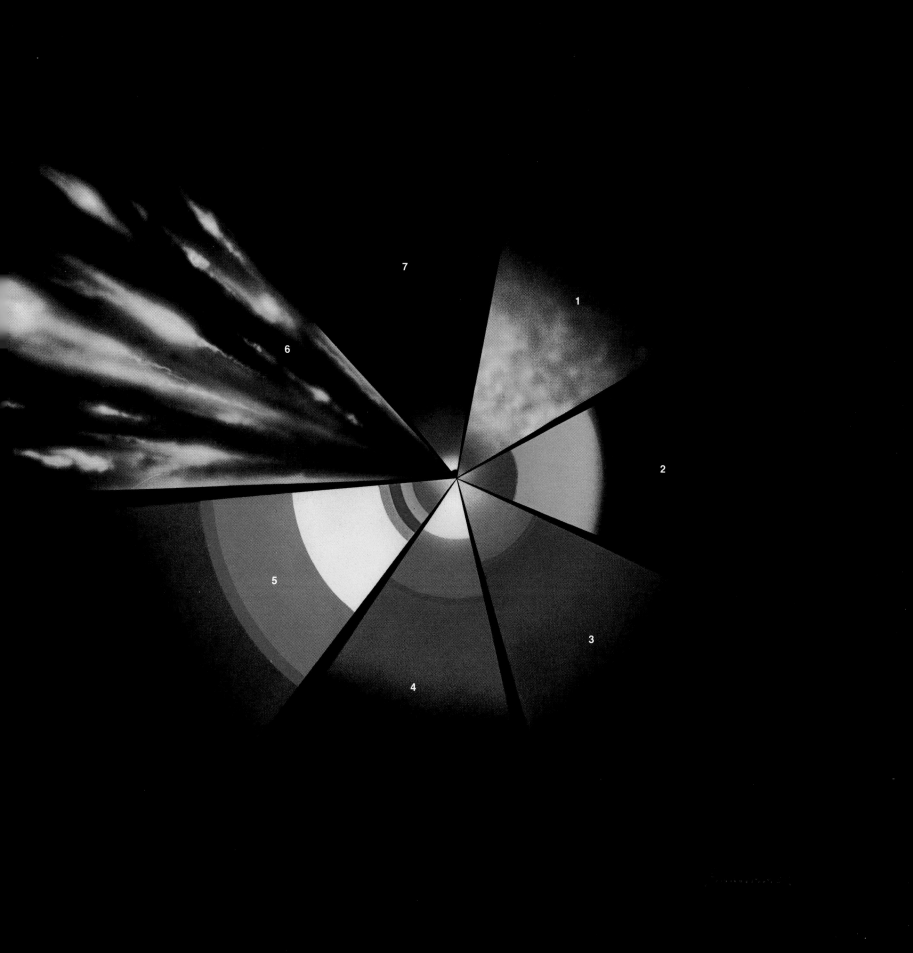

Emergence of the Solar System

1. Supernova blast

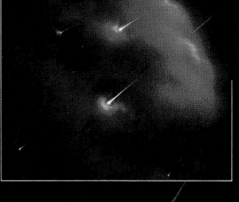

2. Supernova injects new elements into our cloud and initiates its collapse.

3. Random movements cancel themselves in frequent collisions of denser regions.

4. Organization becomes obvious.

5. After the fall of dust into the sun the system becomes clean.

...nets may be condensations
...over from initial conditions.

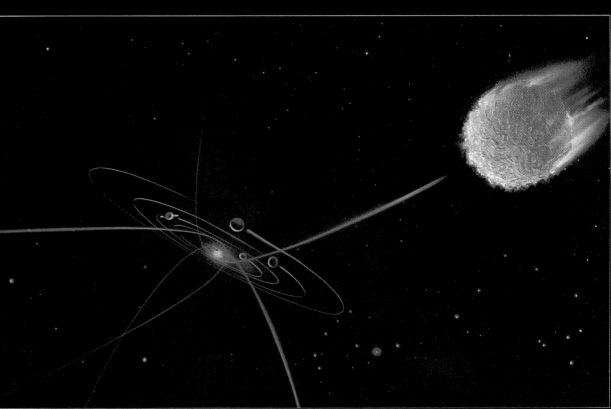

ticles of matter all turning around a center in the same direction—the final net motion with randomness at a minimum.

The distribution of mass within the disk was concentrated in varied rings, some packed with matter, others not. The rings may have been positioned by some gravitational resonance phenomenon in the gaseous disk, for the present arrangement of the planets is very orderly. The mass of the matter in each ring from Neptune's inward must have increased with perhaps a maximum at Jupiter. As Jupiter's protomass collected itself, its concentration of mass and its proximity to the next inner ring prevented a planet's formation there, and thus we have the asteroid belt.

Several large bodies may have formed there, however, to be shattered by subsequent collisions. Meteorites we gather on the Earth are fragments of asteroids. The internal structure of the metallic meteorites shows very large crystals that could only result from very slow cooling over a long period of time. Thus it is expected that they were once insulated by miles of material. Ceres is today the largest asteroid, about 400 miles across. Perhaps it and a few of the other large ones have never suffered the fate of many of their fellows.

Mars, too, may have been too close to Jupiter to collect effectively all the material available to it. The Earth and Venus collected much of the dust in the ring of material within reach of their gravity. The very dense matter had consolidated itself at their centers while the lighter elements such as hydrogen and helium remained far from the central condensations.

Then it happened!

The spherically symmetrical cloud of matter destined to form our star had been slowly condensing toward the center. This accumulation had always been the most massive one in the system, providing the pivot—the center of gravitation that held the protoplanets in their places. When the solar cloud became dense enough, the internal motions of the particles resulted in a crescendo of collisions, each stage more violent than the last. This is described as increasing internal temperature. When the collisions became violent enough to ionize hydrogen—shuck the electrons from the atomic centers, or nuclei—the internal motions became less violent as the energy was diverted to the ionization process. Thus, the cloud lost its supporting pressure, its outward push. The infall of the cloud could not be prevented.

Before it began, the outer boundary of the cloud lay between protoMars and the asteroids. As it all began to fall in, the mass moved faster and faster past the inner planets, sweeping away the lighter, uncommitted stuff that the planets had been slowly drawing to themselves. The already condensed heavy elements formed the cores of such massive planets as Jupiter. The inner planets were core material only.

If Mercury had once been a satellite of Venus, the falling wind could have swept away so much of protoVenus that it lost gravitational control, and Mercury became a planet. This vigorous infall would take only a few months. All that remained for the planets was to make do with what they had. As for the protostar, its sudden accumulation of matter would aid in initiating the internal thermonuclear reactions that led to the birth of our sun.

Some parts of the solar system's primordial cloud did not get caught up in the planet- or sun-forming processes, but condensed in the very cold depths of space. Today we call them comets. Most of these far travelers seem to reside in a huge halo surrounding the planetary system. Only those that chance to approach the sun closely are ever seen by us. No one knows how many cometary bodies are in this halo that reaches halfway to the nearest star. And no one knows when the next one, likely to be bright, will come in for perhaps its first visit in 60 million years.

Our understanding of stellar formation and evolution leads us to believe that the earliest stars—those that began to glow very soon after the galaxy took on its separate identity—possessed few chemical elements heavier than hydrogen or helium. The higher abundance of these elements in the sun, however, shows that the sun is not made of aboriginal matter. Thus it is widely believed that the primordial solar cloud contained atoms that had been manufactured in other stars. The matter of the sun may be having its third stellar experience, or maybe its fourth. This means also that the matter of the Earth, and of you and me, has all been inside stars in the distant past. We are all made of starstuff.

Their orbits swept clean of the nebulosity in which they were born, the planets of the solar system herald their star. Each planet evolved in darkness prior to the sun's ignition; and many gained their final forms only after the light appeared.

H. K. WIMMER

David Morrison

Our Family of Planets

This Voyager I image of Jupiter's Great Red Spot shows details of the famous feature, which may be a wave form in the turbulent Jovian atmosphere.

Only since the 1960s have astronomers begun to appreciate the diversity of the planets and to take the first steps toward deciphering their past. Thanks to the space programs of the United States and the Soviet Union, the 1970s witnessed one of the most dramatic periods of exploration and discovery in human history.

Our explorations, both manned and unmanned, show that the planetary bodies of our solar system include an amazing variety of worlds and environments: gas giants with no solid surfaces at all, objects wreathed in continuous volcanic eruptions, surfaces oven-hot under oppressive acid clouds, cold globes of ice, and one planet with a unique veneer of liquid water. This diversity exists in spite of the common origin of all the planets from the same solar nebula some 4.6 billion years ago. For planets, like stars, evolve, each following its own course.

From the perspective of some hypothetical viewer outside the solar system, the sun has only one companion worthy of notice—Jupiter. Not only is it the largest planet, but Jupiter is larger than all of the other planets and satellites (moons) combined.

Some scientists have claimed that Jupiter is a "star that failed," a body that just barely missed sustaining the nuclear fusion reactions that would make it self-luminous. This is an exaggeration; Jupiter would have to have been many times its present mass to have become even a faint dwarf star. Still, the birth of this giant planet had much in common with the birth of a star. Like the sun, Jupiter accreted out of material in the collapsing solar nebula of gas and dust.

We believe that during its formation Jupiter gained heat as it attracted objects from nearby space. Calculations suggest that the nascent planet's interior temperatures rose to tens of thousands of degrees, enough to vaporize all of the solid material that its gravity had pulled in. Jupiter may have glowed red-hot at its surface, radiating about one percent as much energy as the sun does today. But the central temperatures and pressures never reached a level at which self-sustaining nuclear reactions could begin. Instead, Jupiter's outer layers rapidly cooled, and only deep inside is there a remnant of the primordial heat.

Today, some of Jupiter's birth-heat still leaks to the surface and enters space as infrared radiation. This planet now emits twice as much

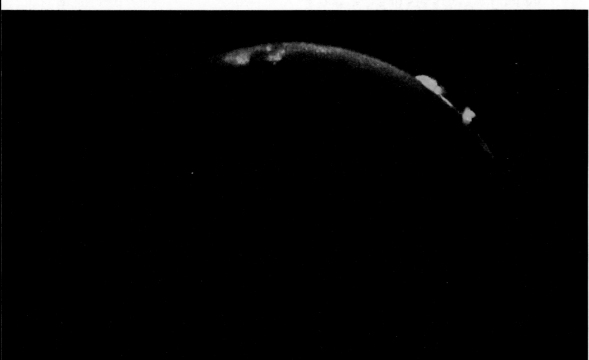

Above: a Voyager I view of Jupiter and two of its moons, Io, left, above the Great Red Spot, and Europa, was taken on February 13, 1979, when the spacecraft was about 20 million kilometers (12 million miles) from the planet. Left, a Voyager 2 close-up of Io, showing violent volcanic eruptions characteristic of this remarkably active moon.

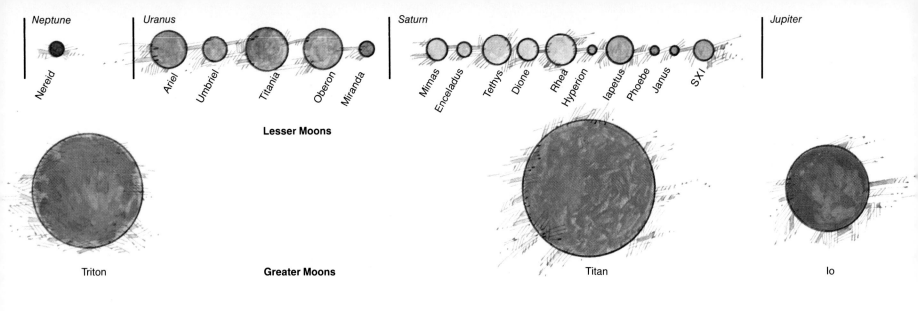

| Neptune | Uranus | Saturn | Jupiter |

Nereid

Ariel · Umbriel · Titania · Oberon · Miranda

Mimas · Enceladus · Tethys · Dione · Rhea · Hyperion · Iapetus · Phoebe · Janus · SXI

Lesser Moons

Triton

Greater Moons

Titan

Io

Moons of the Solar System (with comparison to Earth)

energy as it absorbs from the sun—energy that dramatically influences the structure and motion of the Jovian atmosphere but has little effect on the rest of the solar system.

As a result of Jupiter's strong gravitational pull, no gases—not even light hydrogen—can escape from its atmosphere. Jupiter, therefore, like the sun, retains essentially the same proportion of elements that were in the original solar nebula. Hydrogen and helium comprise about 99 percent of Jupiter. Heavier elements add up to less than one percent. In the outer layers of the atmosphere, these materials are gaseous. Deeper in, where the pressure becomes enormous, hydrogen is compressed into a liquid state. Still farther down the liquid hydrogen assumes some of the properties of a metal. Probably long ago most of the heavier materials sank to the center of the planet where they conceivably exist as a small molten core. As a result of this remarkable composition, and the heat still remaining from Jupiter's birth—with perhaps some heat from continuing gravitational collapse—the planet may be characterized as a sea of gas above a globe of liquid. The whole gigantic ball rotates in just 10 hours—faster than any other major body in the solar system.

A dense white cloud layer of ammonia ice crystals covers most of Jupiter. At some places we see breaks in the ammonia clouds that reveal deeper layers of clouds, including those of both water ice and water. In addition, Jupiter wears a coat of many colors, varying from the brick red of the Great Red Spot to hues of brown, orange, and blue. We do not know their source, but some investigators suspect that these colors may represent organic com-

Above, a chart shows the sizes of the solar system's moons in comparison to the Earth and its moon, pictured at the far right. While our moon is not the largest of the satellites, at one fourth the size of the Earth it is the largest in relation to its planet. The satellite systems of the giant planets, Jupiter and Saturn, resemble miniature solar systems; each may have yet undiscovered minor moons. Left, farthest out of Jupiter's four largest or Galilean satellites, Callisto is also the most heavily cratered. This may indicate that the moon's icy surface has not changed significantly since it was heavily bombarded some four billion years ago by debris left over from the formation of the solar system. Europa, right, smallest of Jupiter's Galilean moons, may have a layer of ice up to 100 kilometers (60 miles) thick over its crust. The great streaks or fractures crazing its otherwise smooth, relatively uncratered surface suggest that its crust has been cracked, perhaps by tidal forces created by Jupiter's powerful gravitation.

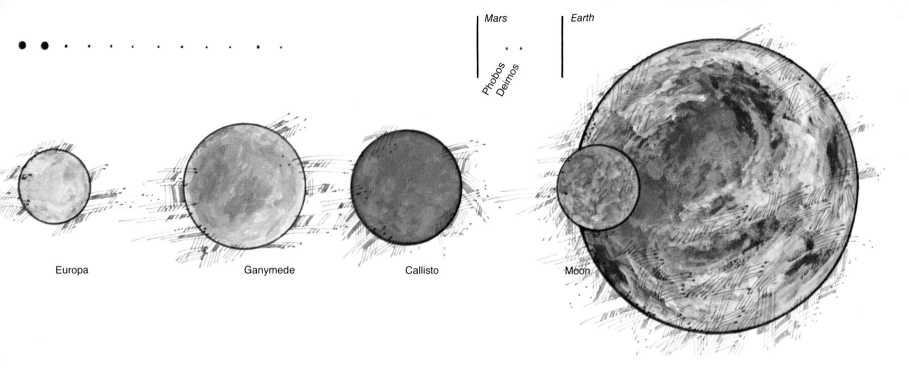

Europa   Ganymede   Callisto   Moon

Mars   Earth

Phobos   Deimos

pounds generated in the Jovian atmosphere by lightning discharges and solar ultraviolet light.

The Great Red Spot is a giant, rotating, high-pressure storm system in the atmosphere, but the mechanisms by which this and other giant storms form remain a mystery. It is larger than the Earth, so large in fact that it was discovered with a crude telescope 300 years ago.

A strong magnetic field extends into

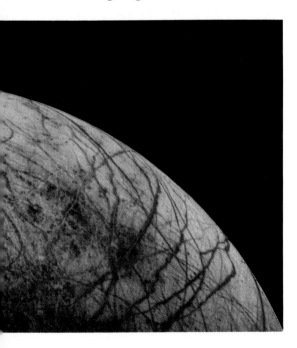

surrounding space to form Jupiter's magnetosphere, the largest feature of the planetary system. The enormous magnetosphere traps charged atomic particles and deflects the solar wind around Jupiter.

Jupiter is the prototype of the four giant, or Jovian planets. The others—Saturn, Uranus, and Neptune—share many of Jupiter's properties, though they are smaller. All are gas bags, consisting primarily of a rather similar hydrogen-helium atmosphere and containing little or no solid material.

Saturn is the next planet out from Jupiter and the next after it in size. Its mass is about 30 percent that of Jupiter, and it rotates a little less rapidly. Saturn also has an internal heat source, presumably a remnant of heating at its birth. The clouds, however, along with the atmospheric dynamics of Saturn, appear to differ from those of Jupiter. Saturn's bland, nearly featureless face lacks Jupiter's large spots and dramatic bands. Perhaps great storms do exist on Saturn, but they lie hidden below a pervasive high-altitude haze.

Saturn's most spectacular features are, of course, its brilliant rings. Little more than a century ago astronomers concluded that the rings consisted of vast numbers of tiny particles rather than solid sheets of matter. Each of

these miniature moons pursues its own separate orbit around the planet, the whole cold squadron circling in a period of a few hours.

There are three chief ring groups made up of numerous smaller ones. Detail from Voyager imagery appears with this essay. In spite of the vast extent of the rings, about 300,000 kilometers (200,000 miles) from edge to edge, they are no thicker than a kilometer or two.

The individual particles of the rings of Saturn are composed of water ice or other frozen material and are fist- to boulder-sized. They can be thought of as snowballs, strikingly different from the rings of Jupiter and Uranus, both of which consist of dark, rocky materials.

Saturn has more than a dozen moons including Titan, believed to be the largest satellite in the solar system until the Voyager coverage of 1980. Most of Saturn's other satellites appear to be icy bodies about a thousand kilometers (600 miles) in diameter, quite unlike Jupiter's moons.

Beyond Saturn lie Uranus and Neptune, both smaller and both apparently containing a higher proportion of rocky and icy material. Neither planet has been visited by spacecraft, and both are so far away that they cannot be seen by the naked eye. Indeed, very little can

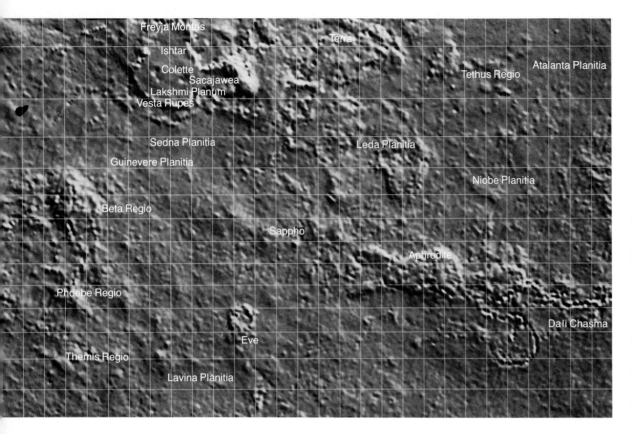

Labels on image: Freyja Montes, Ishtar, Colette, Sacajawea, Lakshmi Planum, Vesta Rupes, Sedna Planitia, Guinevere Planitia, Beta Regio, Phoebe Regio, Themis Regio, Eve, Lavina Planitia, Sappho, Aphrodite, Terra, Tethus Regio, Atalanta Planitia, Leda Planitia, Niobe Planitia, Dali Chasma

Above: the surface of Venus, perpetually hidden from optical scrutiny by thick clouds, was mapped by radar carried on the Pioneer Venus Orbiter. Probes, below, of the dense Venusian atmosphere must be protected against extreme heat and pressure. In contrast, Mercury, far right, has no atmosphere. Heavily cratered and superficially moon-like, Mercury is the densest planet in the system.

be learned about them even from telescopic studies.

Nonetheless three interesting differences between Uranus and Neptune have been discovered. First, Uranus is tipped on its side, with its axis of rotation almost in the plane of its orbit. Neptune—like the other planets—rotates around an axis roughly perpendicular to the plane of its orbit. Second, Uranus has rings, and if there are rings around Neptune, they have yet to be found. The rings of Uranus are different from those of both Jupiter and Saturn in that they are thin ribbons of material only a few tens of kilometers from inner edge to outer edge. And third, Neptune has an internal heat source while Uranus, unlike the other three giant planets, does not. How Uranus escaped the primordial heating experienced by its siblings and why the three known planetary ring systems are so different remain mysteries.

Beyond Neptune lies one more small planet, apparently unrelated to the giant planets. Pluto seems to be more like an asteroid, or minor planet.

With the outer giants to the rear, let us visit the inner planets of the solar system. And very different they are from the outer gas giants. Here, closer to the heat of the sun, the predominant building blocks are rock and metal, with little ice and almost no hydrogen or helium. Hence, the inner planets are solid and dense with at most a thin and tenuous layer of atmosphere.

We will start with our own moon, the largest in relation to its planet. Sun and moon stir Earth's tides and have other important influences that help sustain an exuberant stew of wind, weather, and life.

Mineralogically speaking, the moon is composed primarily of silicate minerals rather like those that make up Earth's own crust. Our moon has little metal, and if a core of iron and nickel exists, it can span no more than a few hundred kilometers. We know our big satellite was formed 4.6 billion years ago by accretion, that it was initially hot and experienced large-scale volcanic activity between 3.3 and 3.9 billion years ago, and that it has been geologically dead for the past three billion years.

The dominant force that has molded the surface of the moon is the impact of innumerable meteoric bodies. The moon's craters come in all sizes, corresponding to the distribution of debris in space, right down to microscopic pits made by the impact of micrometeoroids. With no atmosphere and only the never-ending rain of debris to provide erosion, the moon openly bears the scars of billions of years in the cosmic shooting gallery.

The volcanic activity of the moon was confined largely to an outpouring of fluid lavas to fill in the large impact basins on the side facing the Earth. The lunar lava flows created vast flat plains in the nearside lowlands, which we see as the "Man in the Moon." Apparently the crust is thicker on the far side, for little lava escaped there.

Most of the surface of the moon was formed within the first billion years of its history by impact cratering and volcanism. The moon thus preserves for us the record of an important era of the history of the solar system, an era erased on Earth by our much higher level of erosion and geological activity. But the record of the moon's earliest days was lost in the internal melting and heavy cratering that took place during its first few hundred million years of existence.

Mercury is the innermost planet, circling the sun at just 60 million kilometers or 37 million miles. It also has the greatest density of any of the planets, probably due to a relatively large metallic core. In fact, we might say that it is mainly core. In most respects the geology of Mercury seems to be like that of the moon. The surface is heavily cratered, but evidence exists of widespread lava flows at some time in its early history. Without rock samples we do not know when these events took place. Also, images from the Mariner 10 mission revealed only about half the planet's surface, and there is no telling what surprises may be revealed when, eventually, the rest of the planet is seen at close range. Until then, however, we can think of Mercury as a larger, denser, and hotter version of the moon—an airless, geologically dead planet scarred by its collisions with the orbiting debris of the inner solar system.

Mars is the planet whose surface most resembles that of Earth, and it is the object most extensively studied after the moon, with a total of three flybys, three orbiters, and two lander spacecraft successfully deployed by the National Aeronautics and Space Administration during the past 15 years. Intermediate in size between Mercury and the Earth, Mars is able to hold onto a weak atmosphere, and it has been geologically active, although much less so than the Earth.

The atmosphere of Mars is composed almost entirely of carbon dioxide, with a little nitrogen and argon, and only minute traces of water and oxygen. The surface pressure, however, is only 0.7 percent that of Earth, and the absence of the shielding effect of ozone results in little protection of the surface from solar ultraviolet radiation. Three kinds of clouds sail through the thin atmosphere of Mars. They are composed of ice (like terrestrial cirrus), of frozen carbon dioxide, and of wind-blown dust. The dust clouds can shroud the whole planet during a major wind storm.

Large Martian surface features can be seen even with terrestrial telescopes: light and dark markings that vary with time, and bright white polar caps that wax and wane with the changing Martian seasons. In many respects, the daily and seasonal effects there are Earth-like. The Martian day is just a little over 24 hours long, and the tilt of the axis of rotation is almost exactly the same as ours. Mars, however, takes twice as long to orbit the sun, so the seasons are correspondingly lengthened.

As seen from spacecraft orbiting the planet, the surface of Mars reveals a varied and dramatic geologic history. About half of the surface—primarily in the southern hemisphere—is heavily cratered and looks rather like the moon or Mercury. The craters, though, have been more heavily eroded. In striking contrast, Mars' northern hemisphere is lower than the southern and shows few craters. Instead, vast lava plains predominate, with dune areas and other indications of deposition by atmospheric processes. A series of great canyon systems lies between the two hemispheres. The largest is more than six kilometers (four miles) deep and nearly long enough to span the continental United States. Perhaps most dramatic are the huge volcanic mountains that dot the northern hemisphere. Olympus Mons is the largest, rising about 26 kilometers (16 miles) and spreading more than 600 kilometers (380 miles) across. In all, 20 of these extinct Martian volcanoes are larger than Earth's greatest. Just imagine the drama of Martian volcanism.

Elsewhere, great dry channels rise in the highlands and debouch into the northern plains. Many of these vast Martian channels resemble dry stream

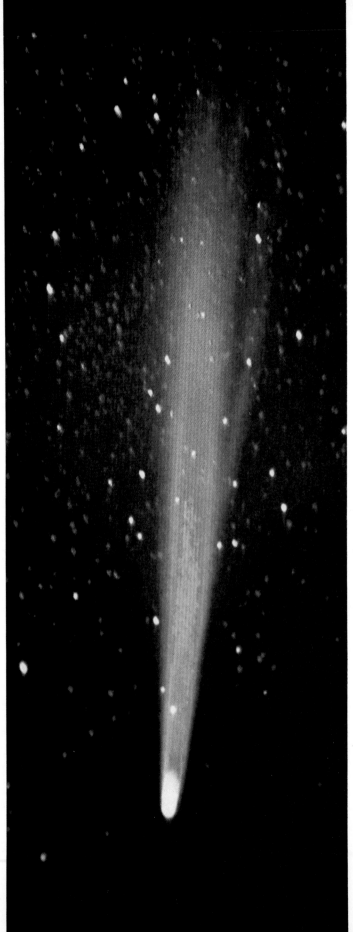

Halley's Comet, right, was photographed in black and white in 1910 during its last visit to the inner solar system. It is shown here after computer enhancing and artificial coloring. This famous comet is due back in 1986 and may be visited by craft from Earth—an important space first. Far right, sunrise over Noctis Labyrinthus on Mars—most Earth-like of the planets. This image was obtained by the Viking I orbiter, illustration below.

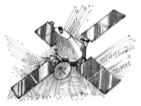

beds; others—even larger in scale—appear to be the result of massive floods. Yet Mars today has very little if any liquid water. If these features were carved by running water then the climate on Mars must once have been much warmer and wetter than it is today. Mars would have undergone climatic changes much more drastic than those which took place during our own Ice Ages.

The bright polar caps of Mars themselves have a dual composition. The seasonally varying deposits are frozen carbon dioxide, or dry ice, that forms a thin coating of frost during the winter months, when the temperature falls below 150°Kelvin (−100°F). But, in addition, there are permanent deposits of water ice, like the terrestrial polar caps. More water in the form of permafrost may lie beneath the frozen Martian surface.

Two U.S. Viking unmanned spacecraft landed on the surface of Mars in 1976. Both found themselves on parched, silent deserts strewn with rocks, under skies tinted pink by suspended particles of red dust. The pervasive red color of Mars is due to iron oxides—rust—in the soil and as a patina on rocks.

A major goal of Viking was to search for life, but the tests conducted at two sites were negative. Scientifically the experiments could not be called conclusive, but it seems likely that Earth alone is inhabited in our solar system.

Venus and Earth—the second and third planets from the sun—are remarkably similar in large-scale properties such as size and mass. Yet they diverged in their evolution and today their properties are fundamentally different, presenting a challenging enigma.

Earth is, as we all know, geologically active, with volcanism, mountain uplifts, earthquakes, and a constant shifting of the crust over a plastic interior. As a result of this geologic activity, and also of erosion by the atmosphere, we quickly lose the scars of meteoric impacts. Few large recognizable craters remain on our planet. Partly for this reason, earlier generations of scientists were slow to accept the impact origin of craters on the moon. Now we know it is Earth, not the moon, that is the more unusual.

Ours is the only planet known to have liquid water, enough, in fact, to cover the whole planet to an average depth of about three kilometers. Also, we alone have substantial oxygen in our atmosphere, because of the constant action of green plants. Just as we have an excess of oxygen, in relation to other planets, so we also have a fortunate shortage of carbon dioxide, most of which is bound up in chemical compounds as carbonates, produced primarily by tiny shelled sea creatures. If these creatures did not exist, our atmosphere would hold a high concentration of carbon dioxide.

Venus possesses a fundamentally different atmosphere. Laden with carbon dioxide, its surface pressure measures 90 times that of Earth, equal to the pressure nearly a kilometer (3000 feet) deep in our oceans; it would require the equivalent of a deep-sea submarine to withstand the atmospheric pressure on Venus. Suspended nearly 50 kilometers (30 miles) above the surface are thick, unbroken hazes of sulfuric acid and sulfur, and perhaps other still unidentified cloud particles. Solar heat, trapped by the carbon dioxide atmosphere and the clouds, generates a powerful greenhouse effect to raise the surface temperature to 740°Kelvin (850°F). Venus resembles nothing so much as hell, complete with heat, a dark oppressive atmosphere, and clouds of brimstone. Somehow Venus has evolved very differently from Earth; we would like to know why.

Relatively little is known about the surface of Venus beneath its blanket of clouds. Radar has been used to penetrate the thick clouds, revealing a relatively flat surface over most of the planet, with a few large continental blocks and mountain ranges. There is evidence of volcanoes and rifts in the surface, but it is not conclusive. Radar has also revealed that Venus rotates

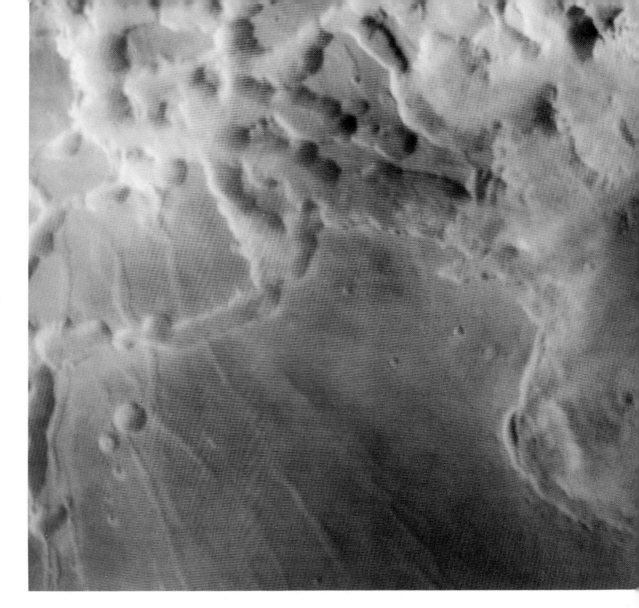

backwards in relation to all the other planets—another mystery.

Since the mid-1960s, the Soviet Union has sent a large number of probes into the atmosphere of Venus. Several of these successfully landed on the surface and broadcast data from that inferno, until they succumbed to the heat. These Venera landers have revealed a desolate, rock-strewn planet, with a surface illuminated dimly through the clouds—a brightness about equivalent to that on Earth on a heavily overcast day.

Having surveyed the major bodies of the solar system, we should now return to the giant planets to take a more particular look at their satellite systems, moons that rival the major

bodies in their diversity. The two Voyager spacecraft flybys of Jupiter in 1979 gave us our first detailed look at the remarkable miniature solar system of the Jovian satellites. The four large objects—ranging from just smaller than our moon to larger than Mercury—were discovered by Galileo in 1610 and are known as the Galilean satellites.

Callisto, the outermost, is cratered, its appearance superficially like Mercury or the moon. Like them, it is geologically dead, and bears only scars from impacts. But the details of the crater forms are unique, as a result of the presence of ice mixed with the soil of the crust.

Ganymede, the largest Galilean sat-

kilometers deep. No one knows.

As a result of strong tides raised by Jupiter, Io, the innermost Galilean satellite, is being heated; only a thin crust exists above a molten interior. As the interior heat seeks escape, great volcanoes burst forth from the surface, sending geysers of sulfur and sulfur dioxide hundreds of kilometers into space. Flows of red, orange, and yellow sulfur cross its surface, making Io the most brilliantly colored of the satellites. The crust is constantly renewed; lakes of liquid sulfur show up as local hot spots. From the great volcanic fountains, sulfur and oxygen ions escape to become trapped in the magnetic field of Jupiter, creating a ring of glowing gas around the planet. Vast electric currents flow between Io and Jupiter, generating radio emissions that can be detected on Earth. Io is the most geologically active object we have ever seen in the solar system.

Through the telescope Saturn's Titan appears larger than the Galilean satellites, but what makes it unique among the moons is its atmosphere, which is more substantial than that of Mars. Clouds blanket this satellite as completely as they do Venus; and until recent Voyager coverage, some people suggested Titan as a possible abode for life. Now it appears to have a dense atmosphere of cold natural gas.

The other large satellites of the outer solar system, including Neptune's Triton, are so faint and distant that almost nothing is known for certain about them, but scientists expect these bodies will prove different both from the objects we already know and from each other. Each world in the sun's family is an individual, and each can play its part in illustrating the history of the entire system.

As each of these bodies—planets and satellites—has developed its own personality, it has obscured the record of its birth. But by extraordinary good fortune, we also have in the solar system fragments of the original condensates of the solar nebula—debris that still bears the stamp of the processes that gave birth to the sun and planets,

Voyager I's high resolution views of Saturn and its moons astonished and delighted astronomers and laymen alike. The image above, taken on November 6, 1980—when the spacecraft was about five million miles away and six days from rendezvous—shows the completely unexpected complexity of the ring structure. Left, the pale bands of the atmosphere can be seen, less sharply defined and apparently less turbulent than those of Jupiter.

ellite, has some cratered areas similar to Callisto. Large areas of the surface, however, show evidence of breaking and slipping of the crust—somewhat reminiscent of similar events on Earth. Ganymede appears dead today, but it clearly had a geologically active history sometime in its distant past.

Europa is smaller and closer to Jupiter. Its crust is made almost entirely of water ice, giving it a bright white appearance. It is covered with a network of light and dark lines which make it look much like an eggshell rolled around until it is covered with cracks. Perhaps the cracking is the result of severe tidal stresses set up by Jupiter's gravitational attraction. Perhaps we are looking at an ice crust floating on a global water ocean tens of

comets, and asteroids.

The asteroids are small planets that circle the sun, predominantly between the orbits of Mars and Jupiter. Their total mass is less than that of the moon, but we believe they are the surviving remnants of the building blocks out of which the rocky inner planets were formed. Many apparently experienced melting and chemical modification in spite of their small size, but the majority are composed of a dark, carbon-rich mineral that many chemists believe to be solid condensate from the primordial cloud.

Comets represent material condensed farther from the sun, where temperatures were low enough for ices to be formed. The solid nucleus of a comet is probably an agglomeration of ices of water and other volatiles, together with dust from the rocky debris—a dirty snowball—the stuff out of which the outer planets and their satellites are thought to have been formed. Apparently the comets were slung out beyond the planetary orbits by the powerful gravitation of Jupiter, and have been kept in the deep freeze of the outer solar system for the past 4.5 billion years. From time to time one of these cosmic fossils falls back into the inner solar system, where sunlight vaporizes some of the ices to form the spectacular head and tail of a comet, permitting us to study samples of the original star stuff from which the planets were formed.

Actual fragments of both comets and asteroids collide with the Earth—meteorites, as we call those that survive passage through the atmosphere. Many scientists believe that the residue of comets is too fragile to survive a fiery entry into the atmosphere.

From the lowliest meteorite or grain of cometary dust to the giant gas planets, all the members of the sun's family share a common origin and have been shaped by similar processes. Some of these processes are now beginning to be understood and might lead us in the not too distant future toward a confident appreciation of our ultimate origins.

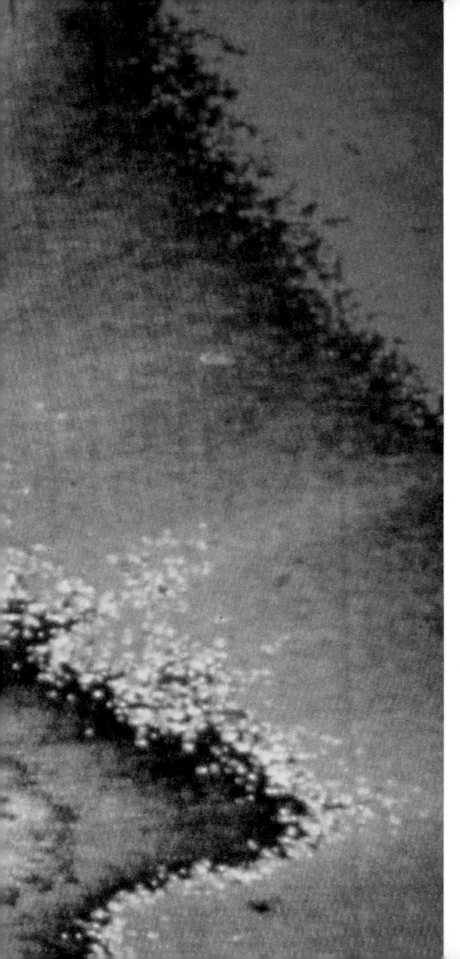

We like to think of the sun as a steady, reliable old friend, wise and stable enough to avoid dangerous youthful excesses of energy. But even our familiar star, at five billion years old just part way through its lifetime, periodically kicks up its heels. Every 11 years, in fact, the sun's activity level builds to a peak.

To study the sun during this active period, the Solar Maximum Mission (SMM) satellite was launched on Valentine's Day, 1980. Since then, SMM has sent back data and images such as the one at left, a computer-enhanced view of the sun's corona, its rarified, four million-degree upper atmosphere, visible from the Earth only during eclipses.

Sophisticated instruments in space and on the ground give astronomers and solar physicists hope that some of the sun's

The Dynamic Sun

remaining enigmas will soon be solved. For there are many questions still unanswered. While most investigators agree on the basic structure of the sun and the nature of its energy production, they find much to disagree over, including the degree to which solar output has varied over the eons. Equally important—and imperfectly understood—are the sun's effects, both short and long term, on the Earth. Worldwide weather phenomena and their possible links to solar variability have thus far proved too complex to be fit into a satisfactory model.

This section opens with our exploration of the inner workings of the sun itself, the thermonuclear furnace that makes it all possible. Professor of meteorology Alistair B. Fraser reveals the visual glories of the sun's interplay with the atmosphere. Plasma physicist George Paulikas looks at the origins—and effects on the Earth—of the solar winds, the invisible storms that sweep through space. Finally, solar physicist John A. Eddy takes us on a journey through present and past to try to find clues to the sun's past and ultimately to its future, and ours.

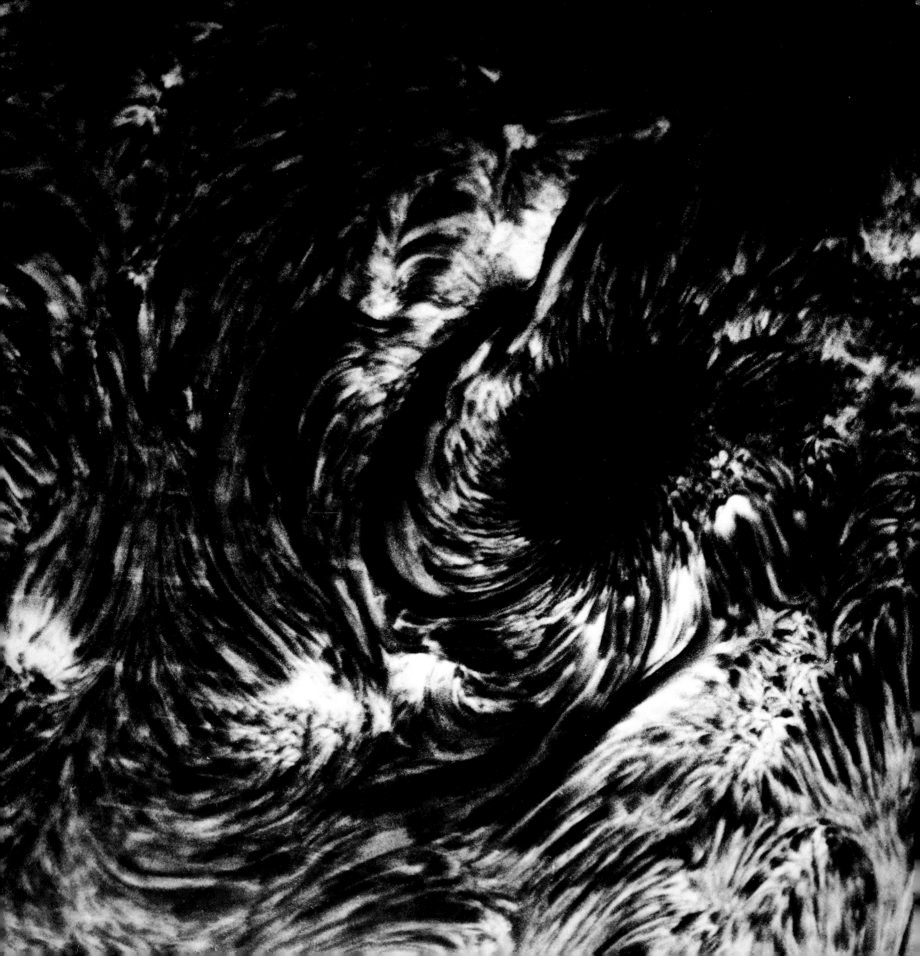

J. Edward Wilzer

The Anatomy of the Sun

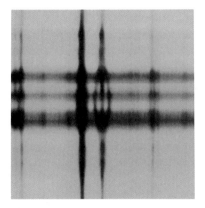

A sunspot as large as Earth darkens the sun's bright surface. Above, spot's magnetic signature, seen as a spectographic line divided into three parts.

Something almost Faustian entered their lives, and many atomic scientists came to feel that they could nearly "touch the face of God." It happened because the nuclear physicists uncovered fundamental knowledge from experiments with particle accelerators (atom smashers) and nuclear devices, and from theoretical advances. And such was the spinoff that today astrophysicists can probe the hidden solar heart.

Investigators, of course, can neither cut away the sun's dazzling mask to reveal the buried springs of power nor dissect the deep roots of surface features. There is no straightforward way to see the sun's interior. So to find out what transpires within, we begin at the outside and dig, feeling beneath the observed surface phenomena. The surgery requires man's most subtle and powerful scientific tools: basic knowledge of conventional physics, and also of how atoms can be expected to behave under extreme conditions. These force matter to spontaneously convert itself into energy within the sun, all other stars, and in thermonuclear devices as well.

Investigators agree on a standard sun model, and say that the bright solar disk represents an organized and integrated energy-conversion system in which matter becomes energy in accord with Einstein's equation of $E = mc^2$. The system appears to be long-lived and self-regulating, functioning through billions of years within the rather close tolerances that we expect from our most reliable resistors, capacitors, and other electronic devices.

Details of an advanced conceptual model appear with the article; and we seek to demonstrate how the parts function together. Let us start our mental slice of the sun at its surface. We say "at" because we cannot say "on." The sun itself consists of gas and has several surfaces, resembling stacked-up cloud layers in our atmosphere. As is also the case with Earth, these outer layers of the sun are very thin compared to the diameter of the underlying globe.

The sun's astrophysical "surface" is the layer from which the light we see emerges; astronomers call it the photosphere. Here the density of the hot gas is so tenuous that on Earth we would consider it a vacuum. But under the force of gravity, the density of the gas increases as one goes toward the center of the sun. And as with any gas on Earth, as the pressure increases so does the temperature. One tenth of the way toward the center of the sun, the density of solar gases reaches that of the Earth's air at sea level. Halfway to the center, the density reaches that of water. At the center, the density climbs to 10 times that of ordinary metals. Though no one has ever measured it, computations suggest that the core reaches 27 million degrees F and experiences a gravitational force 250 billion times that of Earth. And all these constraints apply even as a titanic output of energy works its way outward in the form of motion and radiation.

In such a situation, solar atoms are stripped of their electrons, and in an atom's nucleus the particles can be mashed together. On Earth we have only been able to achieve this "fusion" of atomic nuclei by using atomic devices to drive nuclear particles together in what we call hydrogen bombs. In the sun, atomic nuclei combine or fuse in several ways to produce energy within a tightly controlled and dynamically balanced environment. Scientists gained understanding of these processes only in the 1930s, mainly through the work of Hans A. Bethe in the United States and Carl F. von Weizsäcker in Germany. Technical works have more detail. Here, though, the common reader would be well advised to skim off and sample the cream that rises from the nuclear nitty-gritty. Savor an otherworldly flavor from the realm of the infinitesimal. Also listen: between the lines lies the primal poetry of the universe— worth the effort to catch its crotchety meter and rhyme.

The first, simplest, and most popular explanation of the sun's fusion is known as the proton-proton chain, and involves three steps:

1. Two protons (the nuclei of hydrogen atoms) join to form a deuteron, in the process emitting a positron (an electron with a positive charge)

and a neutrino, a tiny particle that seems to have no electrical charge and little or no mass. Under the conditions that presumably exist in the sun's core, a proton might last for 5 to 14 billion years before being captured by another proton to form a deuteron. There are so many protons in the sun, however, that even this slow rate can produce enough every second to generate the observed energy of the sun.

2. The deuteron then combines with another proton to form a rare variety of helium known as helium-3, with two protons and one neutron, in the process emitting a gamma ray. This part of the chain happens almost instantaneously.

3. Finally, two helium-3 nuclei combine to form the normal helium-4, emitting the two extra protons and another gamma ray. This combination will happen only rarely—once in a million years for any given pair of helium-3 nuclei—but again there are enough of them at any given moment to stoke the atomic furnace.

The net result is that in the core of the sun there is now one helium nucleus where there were once four hydrogen nuclei. But the helium

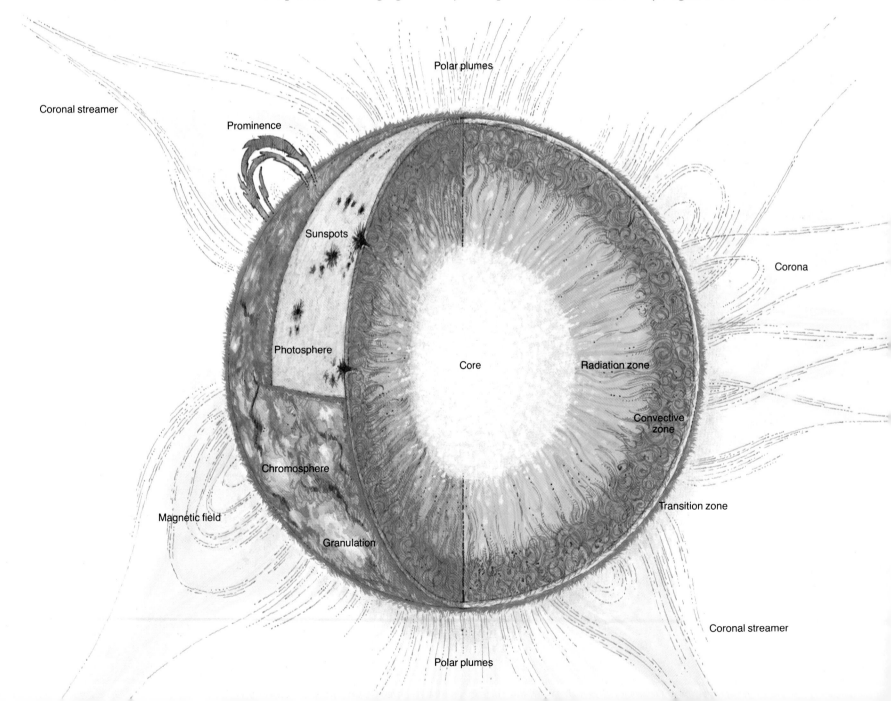

weighs less—seven tenths of one percent less—than the four original hydrogen nuclei. That missing mass has become energy in a nuclear reaction a million times more powerful than a chemical one.

This single proton-proton chain will produce exactly 4.2 one-hundred-thousandths of an erg of electromagnetic energy, which is not very much. Even a whole erg is an incredibly small amount: a one-watt current flowing for one second delivers 10 million. One of the calories counted by dieters equals no less than 40 billion ergs. Thus an individual fusion reaction releases a tiny amount of energy; Einstein's famous formula is at work, but only an infinitesimal amount of mass is turned into energy. What saves us, of course, is that a phenomenal number of these reactions take place. For every gram of mass in the sun, two ergs are radiated every second. And there are 2×10^{33} (two followed by 33 zeros) grams of mass in the sun.

The proton-proton chain can follow several alternate paths. In one of these, a nucleus of boron-8 is eventually formed, only to spontaneously decay into beryllium-8, a positron, and a neutrino. The beryllium isotope then neatly splits into two nuclei of helium-4. Unless the reader possesses considerable specialized knowledge and understanding, all these specifics mean little. But the important thing to keep in mind is that this energy-release chain is of intense interest to solar physicists because the neutrino emitted by the decaying boron is so especially energetic that it, unlike most neutrinos, should be detectable on Earth.

By means of an elaborate experiment involving 100,000 gallons of cleaning fluid at the bottom of a gold mine in South Dakota, physicist Raymond Davis, Jr., has been counting the boron-8 neutrinos for a decade now. And so far there seem to be far fewer than theory would predict. Astronomers are refining their models to account for this, and also seriously considering the possibility that the sun, for the moment, may be in a period of reduced energy generation or perhaps the boron-8 energy circuit may be temporarily switched off.

The second fusion process believed to be at work in the sun, called the carbon-nitrogen cycle, is a good deal more complicated. Just as with the proton-proton chain, though, this other route yields a new helium nucleus and some energy.

Throughout the process, a delicate seesawing can be sensed as one element is transmuted into another and back again. We sketch and highlight the proceedings to suggest the fussy order and quirky economy found in fundamental atomic events. As the name hints, a carbon nucleus combines with a proton to form a nitrogen isotope. The nitrogen decays into a heavier form of carbon than the original, with this nucleus turning right around to absorb a proton and become a heavier form of nitrogen. The new nitrogen isotope combines with another free proton to form an oxygen isotope, which promptly decays into a still heavier form of nitrogen. Finally this nitrogen nucleus absorbs a proton and breaks down into carbon-12, the isotope we started with, and helium-4. Those four protons—hydrogen nuclei—that were absorbed during the six steps have become one helium nucleus. Gamma rays, positrons and neutrinos took flight—a shade less on the energy balances than with the proton-proton chain. Nonetheless, for both cycles the end product weighs a whit less than the original material, a very slight difference that keeps us alive.

The helium by-product is heavier than the predominant hydrogen, and settles to the very center of the sun, around which nuclear fusion continues furiously. Incredibly hot, incredibly dense, this whole region, is what you might call the sun's heart—the deep source of a blaze of gamma rays and fast electrons. Such primitive radiation must age and sweeten before we apprehend it as the light and heat that bathe the Earth.

Equally as important, this tremendous radiation pressure keeps the sun from collapsing. Self-balancing equilibrium keeps it round, as with air inside a soap bubble or a balloon. Should internal radiation diminish, the sun would contract under the force of gravity until the internal pressure grew strong enough to intensify the deep nuclear reactions and produce more radiation. In effect, the sun possesses a thermostat that keeps the energy output rather steady, and also an automatic radiation-pumping mechanism that maintains proper inflation.

Once the vital radiation starts outward from the core, it must penetrate a dense outer shell of hot gas that is opaque to gamma rays, and so such harsh, first-generation energy cannot travel very far. To get the feel of what happens in this region called the radiative zone, think of a gamma ray emitted in the fusion process as a single photon, or bit of light. Before it can move

Opposite: cutaway reveals many solar layers and the sun's unseen interior structure. All energy comes from the core where nuclear reactions are sustained at very high temperatures. Energy seeps slowly upward through an interior radiative zone and then encounters a bubbling layer called the convective zone. The white photosphere that we see with our naked eyes is in reality the top of the convective zone, site of sunspots and granulation. Man's own working model of the sun— an atomic blast—reveals an unstable fireball distorted by intense outpouring of energy, below. The sun itself is more

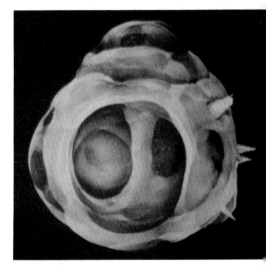

tightly organized, reliably shedding most of its energy as visible light from the photosphere surface. Above it —hotter and thinner—lies the chromosphere, and above that, stretching toward the planets, the wispy yet structured corona. Solar prominences—rooted in photospheric magnetism—thread outward through the chromosphere and arc high into the corona.

very far, it strikes an atom which absorbs it and rises to a higher energy level. The "hot" atom settles back to its previous state after emitting another photon. The energy moves out from the core by successive absorption and emission, over and over and over again.

Imagine people standing shoulder to shoulder across and up and down the entire country, occupying every square inch. Very many letters in the East need to reach the West Coast, and envelopes handed to people on the East Coast may be handed on to people to their west, but may also be handed north or south, and even back in the direction from which they came.

As more and more letters are handed out at the east end of the country, all the letters will gradually move westward as all hands to the east already hold letters. But it will take many steps and a long time before any one letter makes it to the opposite coast.

Such a tortuous process characterizes the radiative zone of the solar interior. Here the tightly packed atoms are too crowded to hold their energy: they collide with each other and pass it on as soon as they obtain it. But when the radiation has worked its way about three quarters of the distance to the surface, it begins to encounter atoms that can absorb great quantities of power.

Energy, streaming out from the core, seems blocked. But gas at this level heats up rapidly and begins to rise to the surface through simple convection. When it reaches the photosphere—the level at which the sun ball becomes transparent—the gas gives up its energy as light and heat and sinks again, going back for more. And now we are right back to the astrophysical skin where we made our initial conceptual incision into the sun. Here the sun's contained, compressed, and thwarted energy can break loose in fireworks.

We view this thin outer layer as a seething, electrically-charged buffer zone of superheated gases, contorted by intense magnetic fields, wracked by the bubbling up of hotter gases from below and thousand-mile-a-second shock waves from explosive flares. The radiation we have been following up from the core passes easily from this turbulent surface through the rest of the solar atmosphere into space. Some eight minutes later about one-billionth of this energy strikes the Earth as sunlight. Other components of the solar output also reach our neighborhood (see pages 100–107).

Above the photosphere lies the chromo-

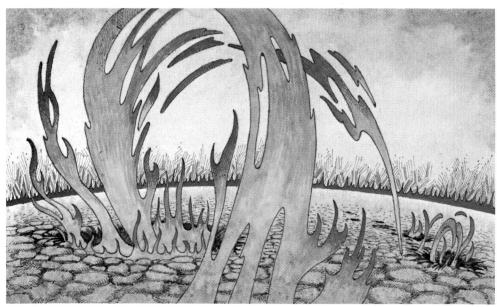

Top: Pic du Midi's clear telescopic eye recorded this tightly packed pattern of granular convective cells. They rise and fall, carrying the energy outward to the photosphere. Each cell is as big as Wyoming. Illustration above magnifies grains, interrupted here and there by the magnetic fields of sunspots. The prominences emerge from sunspot areas and follow curved lines of magnetic force as they arch a million miles above the surface. The sun, a gaseous sphere, seems to rotate faster at the equator than at its poles. An imagined line of sunspots—sprinkled from pole to pole—progressively changes alignment, right. Center view represents displacement after one solar rotation of 27 days; far right, two rotations.

sphere: thinner, hotter, and more dynamic. Pictures taken at very narrow wavelengths show streams of gas, called spicules, shooting up in the chromosphere. These spicules look like giant blades of grass or jets of flame; they last only a few minutes before dying away. Huge convection cells called supergranules rise and sink. Energy seems to be pumped out through the chromosphere and the corona not so much by radiation, but by the motions of the surging granules and supergranules, by pressure waves from convection cells and shock. Finally, the extended outer atmosphere of the sun merges gradually into space. Here in the corona, hot, thin gas is electrically charged and held to the sun by arched magnetic fields. Some of the gas is able to escape the bonds of magnetism and gravity, flowing outward into space as the solar wind.

Almost as important as the corona are the recently discovered holes in the corona. In these large areas, coronal magnetic fields reach straight out into space and channel high-speed streams of solar-wind particles to the Earth and far out beyond the solar system.

The ties between the heart of the sun and its outer atmosphere, are far from understood. We are not at all sure why the sun even has a chromosphere and a corona, or why it shows sunspots, flares, and other dynamic activity. Nor do we know why the lower chromosphere is so cool—about 7000°F, but still hotter than a welder's acetylene flame.

We have discovered from direct observation that the sun's gases are in rapid motion. Not only do they rotate at more than 4000 miles per hour at the surface, they simultaneously move in and out in a convection pattern, just like any fluid heated from below.

The "solar dynamo" theory, the most popular of the current ones that explain sunspots and other activity, postulates that any body of hot gas experiencing both convective and rotational motions will generate magnetic fields, and that solar activity is a consequence of these fields. (The dynamo theory is invoked at other scales: rotation in the Earth's molten core to explain the Earth's magnetic fields, and the rotation of galaxies to explain theirs.) Indeed, magnetic fields are seen on the sun, and are intimately related to sunspots, flares, prominences, and the solar wind.

According to scientists at the Harvard-Smithsonian Center for Astrophysics (CFA), there is reason to believe that the magnetic field not only controls the solar gases but also—perhaps through exotic processes similar to those creating the Earth's radiation belts—heats the gases to the high temperatures we see. If this turns out to be true, then convective heating plays only a secondary role.

Scientists have also discovered that the rate of rotation of the outer layers of the sun varied considerably during certain periods in the past. This suggests an intimate connection between solar rotation and solar activity, consistent with the dynamo theory. But the historical records also hint that the shape and size of the sun may be slowly changing, a result completely unanticipated on the basis of simple theory.

The dynamo model predicts that the solar magnetic field should vary with time, and we do see a 11-year cycle of solar activity. More profound variation occurs as well: the solar activity cycle sometimes fades away for dec-

Bulb-shaped transients in the corona strain against the sun's magnetic fields. Breaking free as they travel outward, looped ends lash out into space.

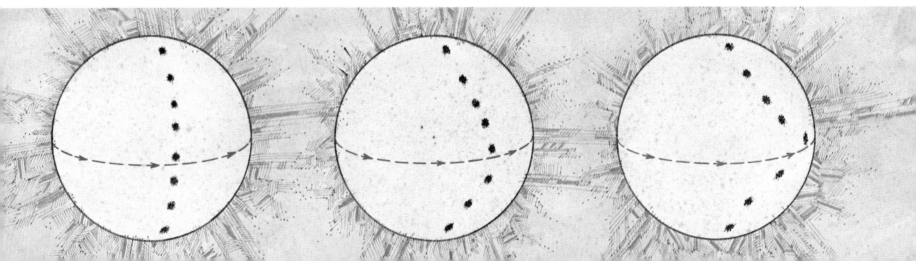

ades, as it did in the "Maunder minimum" during the latter half of the 17th century (see pages 108–115).

Using space observatories to study the ultraviolet and x-ray emissions of other stars, CFA scientists have found that an unexpectedly wide variety of stars are as active as the sun or more so. Unlike our sun, though, some stars, emit a major fraction of their energy in the ultraviolet and x-ray wavelengths, suggesting that their coronas are much more energetic than our sun's, perhaps because of intense magnetic fields brought on by faster rotation.

We can learn about the sun by studying the other stars and vice versa; we can even learn about the sun by studying galaxies. The nuclei of some galaxies can generate huge numbers of ultra high-energy particles. Although the resemblance of these galactic explosions to solar flares has long been noted, only recently have astrophysicists appreciated that the energy source may be the same: magnetic fields that suddenly become unstable. Moreover, the fields may be generated by the dynamo mechanism in both cases. With the sun, convection and rotation supply the required motions. With active galaxies, we may be dealing with a disk of gas orbiting a Black Hole in the nucleus. The disk rotates as it constantly loses gas to the Black Hole. Heat generated by motion in the disk drives convection and, again, the ingredients for a dynamo and time-varying magnetic fields are present. When these fields become unstable, the resulting energy burst can dwarf the power output of even the greatest solar flare.

In any event, it is now clear that phenomena analogous to solar activity are very common among the stars. Indeed, the same nuclear processes that fuel the sun power all the stars. All stars rotate, so we can expect this rotational energy to create magnetic fields which in turn produce cyclical activity. A similar mechanism may be operating on an enormously larger scale to produce violent activity in galactic nuclei and perhaps in quasars.

However far our studies lead, in the end we return to the realities of our particular sun, including its once and future effects on the evolution of Earth and the cargo of life it now carries. And we have more than a glimmer of the sun's ultimate destiny. It seems remarkable, considering that we have studied the sun closely for only 300 years, that scientists are willing to venture a prediction as to the time and even the manner in which our sun will die.

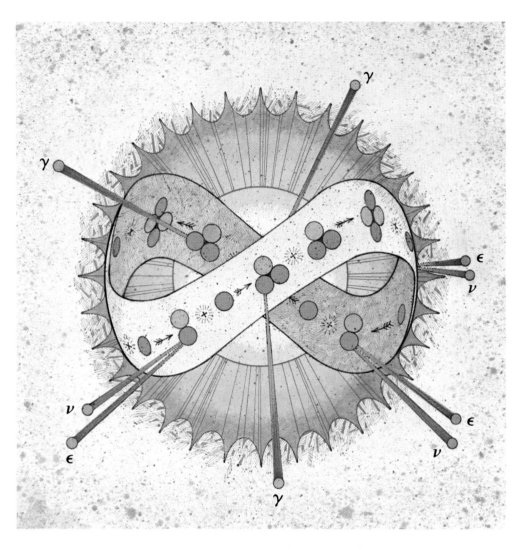

A lot has to do with the family of stars to which our sun belongs, a type of star known to astronomers as a yellow dwarf. It is "burning" 11 billion pounds of hydrogen every second, and has been doing it reliably for the last four or five billion years. The sun is so immense that it has enough hydrogen left to burn for another 100 billion years. In truth it will not. Most of the hydrogen lies near the solar center, far from the outer shell of active thermonuclear reaction. By best estimates, after another five billion years, the available fuel will be used up.

We believe that so far the concentration of

hydrogen in the core has been reduced from 75 to 35 percent; helium has risen from 25 to 65 percent. When the hydrogen in the core is depleted, the core will contract, heating in the process, until helium nuclei can coalesce in groups of three to form carbon and radiant energy. Then the outer regions of the sun will expand past the orbits of Mercury and Venus, making the Earth much too hot for life as the sun becomes a red giant. Still later the sun will contract again, falling inward past its old dimensions until it is a white dwarf, no larger than the Earth. And finally, when every nuclear reaction that can take place has done so, the sun will become dark and dead, a black dwarf. The planets will be equally dark and cold. Any atmosphere left on the Earth will lie frozen on the rocks. One may hope that by then humankind will have found another sun among the stars.

Dr. Alistair B. Fraser

Dazzling Phenomena

"Dazzle mine eyes, or do I see three suns?"
exclaims Edward in Shakespeare's
Henry VI, Part 3. Edward's brother
Richard interprets the appearance as a good
omen for their father's forthcoming battle; but
alas, the day is lost and their father, the Duke of
York, is killed.

From time immemorial people have seen
strange lights in the sky, and, as often as not,
have interpreted them as omens. If Richard and
Edward drew the wrong conclusion, apparently
Constantine the Great (c. 288–337AD) did not.
With a battle pending, he saw a cross over the
sun and proclaimed, "In this sign there is vic-
tory." He won.

If the basis for political or personal omens in
the sky is unsubstantial, the basis for weather
omens stands a bit firmer. A large ring of light
around the sun or the moon has been recognized independently by many cultures as being
symptomatic of impending rain.

All these dazzling phenomena are but sunlight that has been bent either by our atmosphere or by something in it. For the mock
suns witnessed by Edward and Richard, for
Constantine's cross, or the ring that we have all
seen, that something was tiny ice crystals falling out of a cloud. Each crystal acts as a miniature prism that can either reflect or refract the
sunlight to make a spot of light in the sky.
Collectively known as halos, a cornucopia of
spots, arcs, and rings of light trace elaborate
geometric patterns across the sky. Not only the
type and size of the ice crystal, but also the

*Sunlight falling on ice crystals in a cloud bank
produces two sun dogs above a windy North
Dakota snowscape. The effect, a parhelion,
comes about as a sky with settling ice crystals
is touched by sunlight.*

height of the sun above the horizon determines the particular pattern. Some, like the 22-degree halo, are so common that portions of them can be seen by even casual observers and as many as 100 times a year. On the other hand, the circumhorizontal arc, pictured here, appears only within a few weeks of the summer solstice and then just at noontime.

If ice crystals in the air create striking patterns in the sky, so do the water drops of clouds and rain. Sunlight bent by water drops can produce the brilliant rings of a corona, and, of course, rainbows. No matter where you stand, you are at the center of the rainbow's arc which appears before you, the colored edge of a disk of light. And if you move, the bow seems to move. The rainbow is a very personal phenomenon. Each observer sees a slightly different one; indeed, if you carefully examine the bow in the spray from a water sprinkler, two arcs are apparent—one for each eye. Ephemeral behavior by the bow probably accounts for that

symbol of the unattainable, the pot of gold at its end.

Perhaps more than other phenomena of optics, the rainbow has captured the imagination of man. Variously cast as a god, a symbol, and an intellectual challenge, it would seem that almost all cultures have given it a name and embellished it with tradition. Curiously, the Hebraic-Christian tradition of hope runs counter to most others. The ancient Greeks personified the rainbow as Iris, the messenger of the gods, but she carried news of war, trouble, and death. Many African and South American tribes held the rainbow to be a serpent that devoured man and beast, while other people saw it as a giant net poised to capture their souls. Hawaiians call dying "walking the rainbow."

More prosaically though more realistically, our present culture recognizes the rainbow as sunlight that has been bent by raindrops. The light undergoes refraction as it enters, is reflected while inside, and then is refracted

Opposite: over Seattle, Washington, rain droplets in a wave cloud separate sunlight into a blanket of variegated color. And lower, a rare circumhorizontal arc glows in the sky above Blue Glacier on Washington State's Mt. Olympus. It forms high in the sky when the sun is low, gaining its purity of color from oriented ice crystals of hexagonal shape. Mt. Rainier, left, lies within Earth's shadow as first light touches the sky above. At twilight, below, spectral colors suffuse a peaceful sky, the rays being long shadows.

Double rainbow with many faint arcs at Kootenay Lake, British Columbia, left. Below, leafy limb, silhouetted against a sun, is surrounded by the faint colors of a corona. Right: auroral drapery glows. Though derived from the sun, the display is electric. Sunlight falling on the planet's magnetic shield creates an electric potential that sparks the fluorescent show. Opposite: mirage lengthens the legs of pack animals as heated air near the soil acts as a magnifying lens.

again as it exits each drop of rain. A spectrum results, with each color refracted by a different amount to produce the familiar primary bow with red on the outside (closest to the sun). If reflected twice inside a drop, the bent light produces a fainter secondary bow. This inside-out version has the red on the inside. The darkest portion of the sky lies between the two bows. First commented upon in the first century by the Greek scientist Alexander, the region is now known as "Alexander's dark band" (not to be confused with his "ragtime band").

While raindrops are capable of producing a vivid arc, tiny drops such as those found in clouds can create a cloud bow which is nearly pure white, and the two other

colorful phenomena called coronas and glories.

Seen as a series of brilliantly colored rings surrounding the sun (or moon), a corona comes about when uniformly sized water drops diffract the sunlight. This corona is not to be confused with the gaseous envelope outside the sun's chromosphere which goes by the same name. Corona, meaning crown, was applied to the rings of cloud-diffracted light long before the sun's own region of fiery gases was recognized.

Although a corona, at its best, is one of the most spectacularly colored light shows in the sky, few people have ever seen one. The brightness of the sun precludes looking at it directly. Newton found he was able to view it by looking at the reflected image in a puddle on the ground.

A glory, on the other hand, is more subtle in its colors but easier to view. Visible around a shadow when it is cast onto a cloud, one kind of glory is most readily seen from aircraft. In fact, so frequently does it appear around the shadow of a plane that aviators somewhat proprietarily refer to it as the "pilot's bow."

Clouds are not necessary, however, for the creation of striking displays of color in the sky, as testified by the blue of the sky and the colors of twilight. When sunlight encounters air molecules or small dust particles, it scatters in all directions. The blue scattering is so strong that during the day the dome of the sky appears blue. When the sun is low in the sky, both the original sunlight and the blue-rich scattered

97

A green flash, perhaps the most elusive of dazzling phenomena, occurs when the sun lies at the horizon. Air scatters spectral colors, leaving a thin trim of green around the sun where a mirage magnifies it, at left. Below: low horizon effects progressively distort the setting sun as its fading light pushes through the atmosphere. The pilot's bow, or glory, appears when sunlight casts the airplane's shadow onto clouds. Rays pierce cloud layer at right.

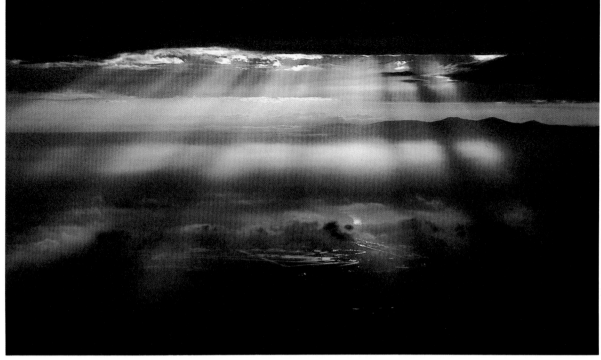

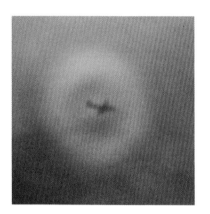

light have a longer path to travel through the atmosphere. The blue is again scattered preferentially, this time allowing other colors to reach us. The various paths that the light can follow through the twilight sky allow the whole spectrum to be seen.

Crepuscular rays, fans of dark lines that cross the twilight sky, are actually the shadows cast by distant clouds or mountains. The lines, which appear to diverge, are in fact parallel, their fan shape being the result of perspective.

As well as scattering the sunlight, our atmosphere can bend it in a much more subtle way. The whole mass of air behaves like a giant lens which alters the shape of distant objects in the scene. This is the origin of mirages, one of which is the wildly distorted

image of the setting or rising sun.

A combination of atmospheric scattering, dispersion, and absorption is also responsible for the abrupt creation of a blue or green rim on the top of the setting (or rising) sun and—on the rare occasions when the whole disk is above the horizon—a red rim on the bottom. Although very thin, the green rim is occasionally magnified by the atmospheric lens so that a momentary —and breathtaking—pulse of green appears on the crest of the sun. This is the "green flash," which in Scottish legend is said to banish all errors in matters of the heart of the viewer fortunate enough to see it. Reason enough to try and espy this most ephemeral of all the dazzling phenomena that grace our sky.

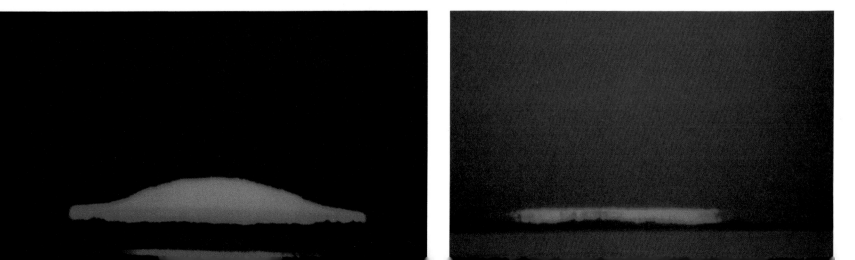

Sun Storms and Solar Winds

George Paulikas

Interplanetary space contains a "weather" of its own with cyclical properties analogous to the seasons on the Earth and with episodes comparable to the violent storms found in our atmosphere. The planets and satellites of our solar system travel through this "space weather" as they orbit the sun. Some, like the Earth, are shielded from the solar blasts by atmospheres and magnetic fields.

Awareness of sun-engendered space weather has come only recently, with the 1970s seeing a big surge in scientific interest. One need only mention August 4, 1972, to bring a smile of fond remembrance to the face of any space scientist. He'll be recalling the big solar flare of that date. Like fine vintages, flares are referred to by year and spoken of with reverence. And not surprisingly, when we consider how very few events with the magnitude of solar flares occur during each 11-year solar cycle.

To be sure, people on Earth cannot see solar flares: they do not brighten the sky. The light from the sun does not change perceptibly. The fact is that the bulk of the energy flow from the sun to the Earth, contained in the visible and near-infrared portions of the electromagnetic spectrum, is constant, stable, and apparently dependable. But viewed at shorter wavelengths, in the ultraviolet and x-ray region of the spectrum, as well as at very long radio frequencies, the sun is indeed a variable star. And flares broadcast their energy in such invisible modes of the electromagnetic spectrum.

The fractional variability of the whole solar output seems very small indeed; it is only a fraction of one percent of the total. Yet this seemingly insignificant energy output helps cause variations in the Earth's ionosphere, with important practical effects on radio communications. And in addition to energy, matter is also shed by the sun.

So space in the solar system is never a simple or perfect vacuum. Even when quiet, the sun continually boils off the outer layer of its atmosphere, spewing it into space. This outwardly flowing material—the solar corona—is ionized, and the flow of such an ionized gas, or plasma, carries the outer fringes of the solar magnetic field off into space.

In this manner, the space between the planets is washed by a magnetized plasma—the solar wind—which streams in all directions away from the sun. Fast by terrestrial standards, the solar wind moves past us at speeds ranging from 300 to more than 700 kilometers (200 to 400 miles) per second with the higher velocity streams squirting forth from orifices in the solar atmosphere, called "coronal holes." The name originated from the localized absences of the x-ray emission which characterizes the corona—a lack which appears to be related to the configuration of the solar magnetic field in these regions. Such loopholes allow the ready escape of high-speed solar wind into interplanetary space.

The effects of the solar wind can actually be seen—in the streaming "hair" of comet tails. Often the cometary tail, blown out from the direction of the sun, is split in two parts: one part deflected by sunlight, the other by streaming solar wind.

The solar wind also interacts with the magnetic fields of other planets to form the force shields that we call magnetospheres. Spacecraft reconnaissance reveals that Mercury, Jupiter, and Saturn possess these features. Venus and Mars, however, apparently lack fields strong enough to form them. Such optically invisible shields occur even when the sun is free of sunspots and no eruptions or explosions are occurring in the solar atmosphere.

In some way not yet clearly understood, all of

As seen by an x-ray telescope aboard Skylab, the sun is far from uniform and symmetrical. Areas of activity (blue and purple) show up in the very hot corona. Various loops, arches, and arcades of hot, thin gas constitute the corona, and these features often stream out into space. Where x-ray emission is low, some relatively dark regions appear. These coronal holes are believed to be the major source of the solar wind that buffets Earth and other planets.

Illustrations by Jim Lamb

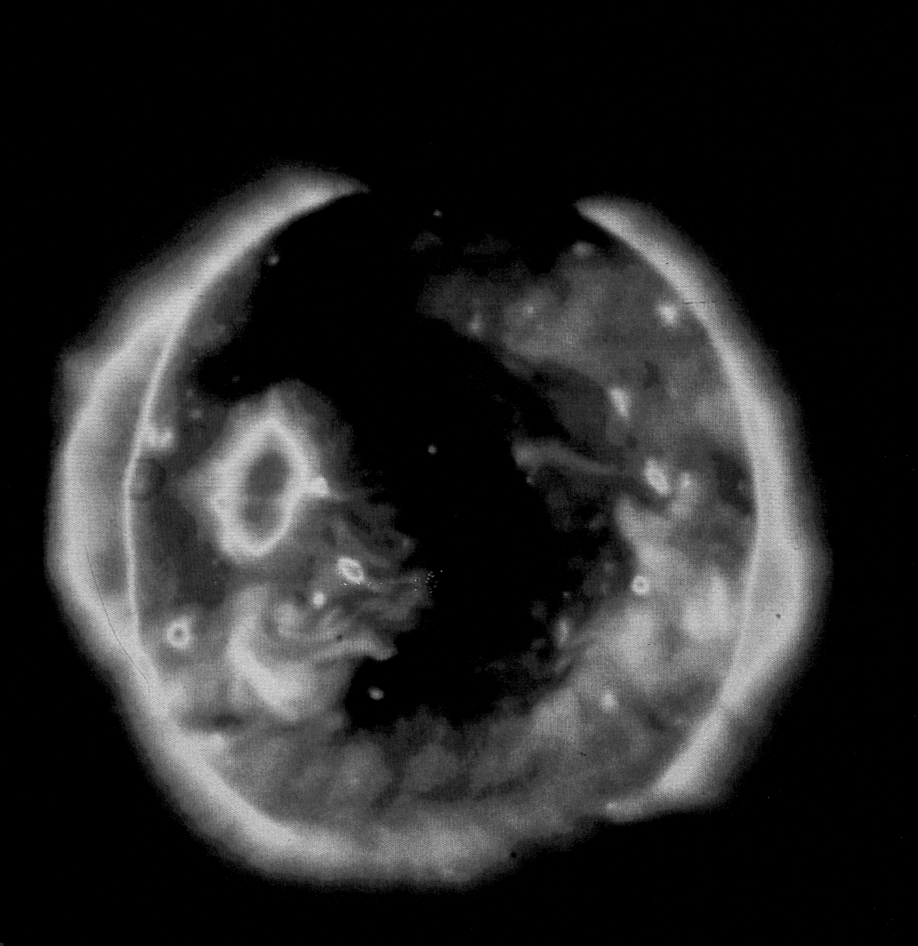

A schematic approximation —as someone in Galileo's time might have pictured it— illustrates the effect of the solar wind on Earth's extended magnetic field. Magnetic lines of force that face the sun during daylight are compressed by the flow of solar wind. On the night side, the flow elongates the Earth's field to distances well past the moon. The region of nearby space into which Earth's magnetic field is confined is called the magnetosphere.

the changeable features of the sun on both the large and small scales are connected with solar magnetic fields. These variations include the extension of the sun's magnetic field into interplanetary space and the merging of the interplanetary field with the Earth's own magnetic field. The solar influence distorts Earth's field into a complex, comet-like shape, with Earth's magnetic "tail" extending away from the sun well past the orbit of the moon. Particles trapped in our planet's magnetosphere form bands of radiation called Van Allen belts after their discoverer.

Mercury possesses a mini-magnetosphere, barely extending one planetary radius away from the surface of the planet. On the other hand, Jupiter's magnetosphere is a true monster. If you could see it in the night sky, it would loom as large as the full moon. After the sun itself, the Jovian magnetosphere is the largest single feature of the solar system.

The flow of the solar wind extends a considerable distance in all directions outward from the sun's sphere. Our interplanetary spacecraft, such as Pioneer 10 and 11, are presently still within the heliosphere, the domain of the solar wind, but they will eventually leave the solar

system. We expect that ultimately the outrushing solar wind must collide with the interstellar medium, but such a frontier has not yet been found.

It may well be that at 50 A.U. (astronomical units: one unit equals the mean distance between Earth and sun or 93 million miles) we would encounter remnants of the interstellar medium, the vestiges of stellar winds which have slowed down. These particles from suns other than our own hit our solar wind. The pressures may cancel out, and the once moving star materials may stall in space.

The most energetic particles reaching the heliosphere frontier would have originated in that most spectacular manifestation of solar activity, the solar flare. We mentioned its awesome power at the beginning of this article. A flare (so-called because this phenomenon was first observed as a spatially localized brightening of some lines of the solar spectrum) is a stupendous explosion in the solar atmosphere. It typically begins in a small region containing complicated magnetic field structures near, but not in, sunspots. The volume containing the initial explosion is shaped like an arch; but it may spread rapidly, exploding into an arcade—

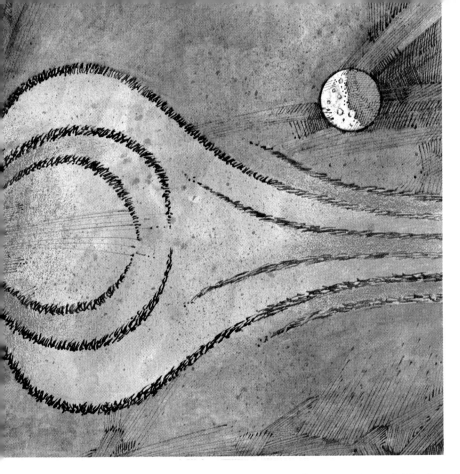

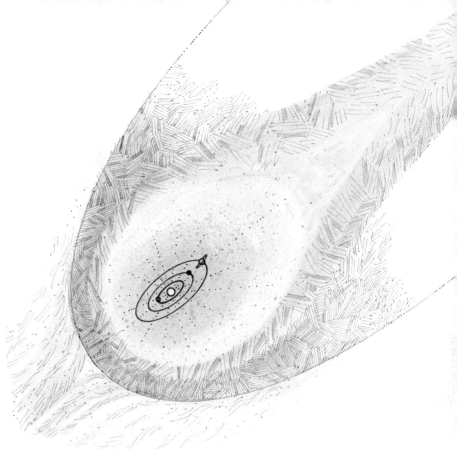

a tunnel of arches—soon engulfing more than one ten-thousandth of the solar atmosphere. Electrons and protons are somehow accelerated by the flare to velocities approaching that of light. These particles burst violently into space.

Away from the protective—although leaky—shield of the terrestrial magnetic field, such sun-produced cosmic rays are a real menace to spacecraft and their inhabitants. Even vehicles orbiting close to the Earth can be hurt, as holes in Earth's magnetic shield allow the invasion of solar cosmic rays at the high latitudes. A flare's electromagnetic radiations require only 8.5 minutes to complete the sun-Earth transit. The smart way to avoid or minimize the annoying or troublesome effects of this flare radiation is to predict the event—such predictions are, at present, about equal measures of art and science.

A solar flare also blows a portion of the solar atmosphere into interplanetary space, creating a much enhanced solar wind. This sun material moves slowly, and only after several days does it reach Earth, when magnetic storms, changes in the Earth's radiation belts, and spectacular auroral displays occur.

Effects of solar flare and magnetic storm eventually penetrate Earth's magnetic shield,

invariably degrading certain kinds of long-distance radio communications. Indeed, in the old days, telegraphs clicked away mysteriously with no human at the key. Even modern cable transmission and reception may suffer, particularly those circuits lying at higher latitudes. Currents induced in these long conductors may upset switching networks.

But it is in space that the brute force of a solar flare is felt. Radiation dosages may increase by factors of a thousand or more. Truly dangerous flares are rare, occurring perhaps a few times during an 11-year solar activity cycle. Yet precautions must be taken if reliable protection of men and electronic equipment in space is to be achieved. Manned spacecraft, for instance, are immediately alerted about the potentially hazardous solar incursions. Similarly, high-flying aircraft traversing the polar regions routinely receive warnings as solar storms reach the planet.

Warning broadcasts come from space-environment forecast centers, notably those operated domestically by the National Oceanic and Atmospheric Administration at Boulder, Colorado, and the U.S. Air Force Global Weather Central at Offutt Air Force Base in Nebraska.

Solar wind, expanding spherically from the sun, fills space beyond the outer planets. The volume of this interstellar space is called the heliosphere and it moves through space with the entire solar system. The solar wind on the leading side presses against the interstellar medium—far outer space filled with particles shed from other stars. As it moves, the heliosphere grows rounded on one end, and trails off at the other.

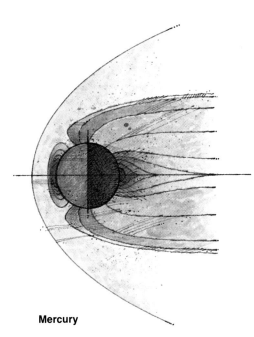

Mercury

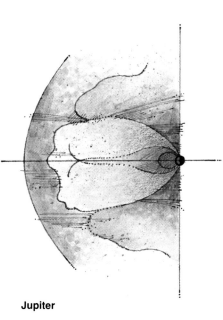

Earth

Jupiter

Left: a cutaway of Earth's magnetosphere reveals its outer and inner configurations. A shock wave is formed where the magnetosphere encounters the solar wind. The planet's compressed magnetic field lies closer to Earth. Various shadings reveal the many complicated domains of magnetic space plasmas—fields of energetic particles from the sun. The region closest to the Earth holds the famous Van Allen radiation belts. Top, from left to right, appear the magnetospheres of Mercury, Earth, and Jupiter. Mercury's shock wave lies close to the surface, while Earth deploys a larger buffer, and great Jupiter puts out an immense magnetic umbrella against sun storms.

The centers provide short-range forecasts as flare phenomena unfold, as well as longer-range predictions—a variety of solar-activity alerts ranging from long-range solar flare warnings to predictions of ionospheric effects and magnetic fluctuations at the Earth's surface. Scientists, commercial customers, and agencies of the government receive the forecasts.

Forecast centers deal with dramatic solar phenomena which batter the planetary environment, yet the sun's day-to-day fluctuation, though more subtle, may have the greater cumulative impact.

The density of the solar-wind plasma near Earth is typically very low—a few ions and electrons per cubic centimeter. The magnetic field imbedded in the plasma (the interplanetary magnetic field) is only one ten-thousandth of the strength of the magnetic field of the Earth's surface. Yet all forms of solar activity which reach the magnetosphere manifest themselves ultimately in the atmosphere or ionosphere. Very energetic particles more or less permanently trapped in Earth's Van Allen belts are the final product of a complex series of interactions of the solar wind with the Earth's magnetic field, the ionosphere, and the neutral atmosphere. Results can be dramatic.

Skylab is falling! went the cry in 1979. And just as all objects in orbit at altitudes less than 700 kilometers (400 miles) are destined to return to Earth within a human lifetime, so Skylab had its downfall decreed. Drag exerted on orbiting bodies by the tenuous upper atmos-

phere brings them down. Skylab's early demise was due to higher than anticipated atmospheric drag. The density of the upper atmosphere, and hence the pull exerted on satellites, depends on solar activity, a relationship first described in detail by Smithsonian scientist Luigi G. Jacchia.

Estimates of satellite lifetimes involve prediction first of solar activity and second of the air density changes associated with such activity. Foretelling changes in space weather—and the resulting effects on atmospheric density—suffers from frustration analogous to that faced by your friendly National Weather Service forecaster. But like NWS, we're improving fast.

In terms of prognostication, no area of research on solar activity has generated as much controversy and speculation in recent times as the question of the possible connection between solar activity and weather and climate at the Earth's surface. Scientific controversy has percolated into the popular press, as the opinions of proponents and opponents of varied ideas regularly make the news. The question is clearly of the greatest interest and importance, as it should be, for the entire agricultural and energy supply system of the world appears only marginally stable. Imagine the Corn Belt slipping 100 miles south and leaving all those Iowa farmers in the lurch. The ability to predict changes in the weather and climate, based on the known regularities in the output of solar plasmas, magnetic fields, and ultraviolet radiation, could be an important factor in economic planning and resource allocation on a regional, national, and global scale.

The evidence that has been presented in support of the hypothesis that the variable sun affects the weather and climate usually has been rather incomplete. The opponents of the hypothesis legitimately question the validity of the information used in the studies and the persistence of the correlations as well. And even the proponents admit that they do not have a firm handle on the problem. Even such fundamental questions as to whether climate changes as a result of long-term variations in solar luminosity or alterations of other solar emissions have not been satisfactorily formulated. We are thus left with the unhappy situation that the potentially premier impact of the variable sun on society and technology cannot be readily addressed with present knowledge.

Fortunately, one effect on Earth of interplanetary weather has few economic implications, and is awesome and benign at the same time—even pretty. I speak of Earth's auroras. Zones of auroral display encircle the magnetic poles in both hemispheres. In the Northern Hemisphere the auroral zone passes through central Alaska, dips down nearly to the U.S.-Canadian border in the longitudes near Minnesota, then swings northward again through the center of Greenland, northern Norway, and the northern sections of the Soviet Union. The optical emissions which are the auroras are produced by the impact of electrons and protons raining down on the Earth's atmosphere from the magnetosphere. The phenomenon—spatial and temporal—of the aurora is incredibly complex and not well understood. The rapidly changing display in the night sky is a cloak of light and color—unvarying in its fascination—shifting draperies and swirls, diffuse arcs and rays.

Imagine the impact of the aurora on ancient peoples as they watched the heavens which changed from fiery red to green and white arcs moving silently across the sky. The possibilities for the development of an inventive mythology for the people living at high latitudes based on their experiencing the aurora would seem almost limitless. And indeed, saga-spinning Eskimos and American Indians, as well as Scots and Scandinavians, live up to expectations. In the Southern Hemisphere, the Maoris were aware of the aurora and believed it due to fires, lit by their ancestors whose canoes had drifted off into the Antarctic.

Although auroras are usually confined to the austral and boreal realms, occasionally they extend much closer to the equator because of great disturbances in the magnetosphere caused by the solar wind. In 1958 a spectacular aurora covered much of the Midwest, giving the appearance of the reflection of a conflagration that reached from horizon to horizon. Fire companies of several midwestern towns drove north in search of the nonexistent fire. An aurora is said to have been observed in 1902 in Jakarta, Indonesia—practically on the magnetic equator—as far as possible from Earth's end zones.

As visually pleasing as the aurora may be, the presence of this luminosity is inevitably accompanied by radio interference. Auroral effects cause the ionosphere to scintillate, and radio waves traversing the ionosphere can thus also shimmer. Even so, observations of such scintillation of radio transmissions and other related effects have helped us amass knowledge quickly. In a few decades our perceptions of the influence of the sun on interplanetary space and on the planets have changed dramatically.

Only yesterday interplanetary space was perceived as pristine, disturbed occasionally by a vagrant cosmic ray. We know today that space to the farthest reaches of the solar system is filled with the expanding remnants of the solar atmosphere. All of the planets and their satellites sail on their regular paths through well-weathered space. At this very time, our interplanetary spacecraft are heading toward the outermost reaches of the solar system, exploring the farthest domains of the sun. These spacecraft may—or may not—find the boundary of the heliosphere before distance overwhelms their even now feeble signals.

It is an exciting time to be alive. To our grandparents this essay would have seemed hardly credible. Imagine: going to the planets! To our grandchildren, this essay will definitely be old stuff. *Sic transit.*

Night photography by satellite reveals auroral lights above southern Canada. The bands and pools of light appear broken because three orbital passes were required to photograph the North American continent. In this three-hour period the aurora changed but the city lights remained the same. Here they outline United States shores. Northern and southern lights occur because sun-energized electrons rain down from space. Big fluorescent features generally occur close to the polar circles, and shine brightly in mid-Canada, Scandinavia, and in the Antarctic.

John A. Eddy

The Search for Solar History

Immissione Refractoria composita.

What has happened will happen again,
and what has been done will be done again,
and there is nothing new under the Sun.
Ecclesiastes 1:9

And what of the sun itself? Is it the same from day to day with nothing new? Or has it changed? What will it be like a hundred years from now, or a thousand? Is the sun the same today as when it shone down on the head of Shakespeare, or Tutankhamun, or Cro-Magnon man?

The search for solar history and the study of possible long-term changes in the behavior of the sun are questions of real interest in science. We can be fairly sure, from the start, that the sun has never varied wildly or erratically. Our sun belongs to a class of middle-aged stars that live in their slowest stage of stellar evolution, with the promise of a long, full life ahead. Other stars of the same type—some younger and some older—are steadfast in their brightness and well behaved, so far as we can tell.

This has surely been our impression of the sun. Daily it rises and daily it sets, shining always the same. Because of its predictable apparent movement and steady brilliance the sun has become the emblem of reliability and constancy. This common notion, based on perceptions more poetic than precise, may reflect, however, little more than an awareness of our utter dependence on the sun, and the natural desire for regularity and stability.

Solar Variability

We have known for centuries that the surface of the sun is pitted with dark spots that come and go. That is, it is imperfect, and it changes. Observations have shown that sunspots are but a minor symptom of more dynamic and pervasive variations in the solar atmosphere. Our sun *is* a variable star, in the sense that its burning surface is never tranquil: its outer atmosphere is torn by a succession of impulsive events such as solar flares and coronal eruptions. How different would be our everyday concept of the sun were we close enough to hear it roar!

The best-studied solar changes obey a seemingly ordered pattern, with the number of sunspots and associated events increasing from a minimum to a maximum and back to a minimum again in a definite cycle of about 11 years. The discovery of this regular fluctuation in the mid-1800s heralded the dawn of the modern study of solar physics. At about the same time the photographic plate and the spectroscope came into use. Thus we may say that the intensive study of the sun, and hence our best documented knowledge of its behavior, began in about 1850. Earlier than that, as we shall see, our view looking backward into solar history is far less clear, as though a curtain fell.

A fluctuation in addition to the 11-year cycle has also been watched since around 1850—a gradual drop to low sunspot counts about the turn of the century, followed by a rise to a broad maximum in the mid-1900s. Throughout this gradual 20th-century rise the number of sunspots continued to follow the 11-year cycle. But for about 60 years the sun marched to a different drummer as well. During the same time, disturbing effects of the sun on the Earth's magnetic field gradually increased. The aurora borealis and australis, the Northern and Southern lights, burned brighter than usual. Although we lack continuous measurements, we can be reasonably sure that the intensity of ultraviolet and x-ray radiation from the sun followed a similar, long-term trend.

Sources Of Solar History

For the time before 1850, records of diverse kinds have been discovered, though, from which we can recover a part of the hidden earlier history of the sun, and in recent years a growing effort has gone into their interpretation. To our surprise as we look backward we find evidence that challenges mankind's original and nonscientific assumption of solar regularity.

Most valuable are observatory records and historical accounts. From these, for example, we can construct an almost continuous record of sunspots to the early 17th century. As we press backward, the quality of these records diminishes, mostly because of gaps in the record. The oldest accounts are quaint and often in Latin, the language of Renaissance learning and scientific discourse; but much of what they contain is still valuable today. At about 1610, however, we come against a second curtain that restricts our view: before then the telescope was not available. Any observations of the sun from before that limiting date were made of necessity with the unaided eye and are therefore less precise.

Telescopic Records

In the 1850s Rudolph Wolf, an astronomer at Zurich with an interest in history made an extensive study of early telescopic reports of spots on the sun, the simplest index of solar behavior. His work provides our best and most

Preparing the famed sunspot compendium of 1630, Rosa Ursina, Jesuits working with Christopher Schiner trace solar images projected by their telescope, opposite. Below, the bear cachet honors the work's sponsor, the Duke of Orsini, his name being derived from ursa, *Latin for "bear." The rose symbolizes the sun and perhaps the Church in which Schiner was a priest. Opposite, the duke himself basks in the radiance of roses and suns, the suns with spots accurately shown during the course of a year.*

detailed source of solar history: an invaluable diary of the number of sunspots that now extends 370 years into the past. From about 1700 we see the clear imprint of the modern sunspot cycle.

Then as now, the sunspot period averages slightly more than 11 years. The number of spots reported at the maximum of any cycle varies considerably, however, and we can pick out longer-term trends like our 20th-century rise. We also see a prolonged suppression of sunspots between about 1800 and 1830, and a subsequent recovery that resembles the sun's behavior in the present century.

The earliest parts of Wolf's reconstructions are the least reliable, and this led him to terminate the published listing where the trail of early sunspots seemed to run out about 1700. We now believe that he stopped at the most interesting part—a span of about 70 years when sunspot activity fell into a dramatic decline.

Solar inactivity persisted between about 1645 and 1715, coincident fortuitously with the reign in France of Louis XIV, *le Roi Soleil*—the Sun King. This curious chapter in the life of the sun is, we now think, the most remarkable feature of its recent history. It has also become a clue that has guided us to a series of new discoveries.

Astronomers of the 17th- and 18th-centuries, who made the records that Wolf was working with, had themselves observed the 70-year absence of sunspots. Some went so far as to tie it to possible terrestrial effects, and until the mid-1800s the unusual absence was commonly mentioned in astronomy texts. With the discovery of the 11-year sunspot cycle at mid-century, however, the notion of a regularly fluctuating sun was so fully adopted that any observation of irregular behavior—particularly when it was based on accounts then nearly two centuries old—was easily dismissed.

In 1887 the German astronomer Gustav Spörer once again called attention to the unusual absence of sunspots between 1645 and 1715. Walter Maunder, chief of the solar division at the Royal Greenwich Observatory in England, also took up the case, and during 30 years pressed the search into the original records, convincing himself that the phenomenon was real. At the time, pleas to recognize the period as a significant anomaly in the life of the sun went largely ignored.

Recent studies, invoking evidence and techniques unavailable to Spörer or Maunder, have confirmed all their early conclusions. In fact, the period of the quiet sun between 1645 and 1715 has now been named the Maunder minimum. Comparisons with the modern record of world climate, also inaccessible to Maunder, have raised the possibility that the Maunder minimum of suppressed solar activity may have been related to a coincident climate anomaly—the prolonged cold of the so-called Little Ice Age. Because of the scarcity of meteorological data, the climatic connection is less certain.

The telescope came to astronomy barely in time to catch the Maunder minimum; what sketchy information we have of the sun between 1609 and 1645 shows a gradual fall in sunspot numbers. During the early 17th century we can also trace a coincident change in the way the sun rotated.

The next step is to peer behind the curtain of mystery that hides solar activity before 1610. Have there been earlier periods of a quiet sun? How common, or uncommon, were they in the longer history of the star?

Pre-telescope Accounts

For earlier centuries, we must turn to the recorded histories of the Orient. In early China, Japan, and Korea unusual features of the sky, by day or night, were reported in dynastic or local annals that still survive today. Dark features on the face of the sun were also recorded, typically in picturesque terms: "within the sun there was a black spot . . . like a hen's egg"—and with it, the date of observation. From these unlikely sources have come valuable information on the history of the sun. From a catalog recently prepared from these naked-eye observations, two more periods of low sunspot activity have been perceived. That between 1400 and 1600 is called the Spörer minimum, an earlier interval between 1280 and 1350, the Wolf minimum.

The Aurora

Another, better source of early solar history are reports of the aurora from diaries, legends, and lore. The aurora is a direct result of events on the sun, which in turn relate to the number of sunspots (see pages 100–107). And since the Northern or Southern lights are spectacular and easily seen without a telescope, we have a long though surely incomplete record of their occurrence—in carefully compiled catalogs back to the 5th century B.C. That record can be read,

Opposite: stream trickles down an Alaskan valley that was originally gouged into its rounded profile by an irresistible river of ice. At right, an observer ponders possible links between solar variability and the onset of historic glacial advances. Above: the so-called Little Ice Age from about 1500 to 1850 lives on in a famous scene painted by Pieter Bruegel the Elder. During the prolonged period the climate of Northern Europe was much colder than it is today. Astronomical sources reveal a conspicuous absence of sunspot activity at this time. The coincidence of this solar Maunder minimum with the Little Ice Age may indicate a correlation between Earth's climate and sun behavior.

with certain reservations, as a proxy history of sunspots.

In auroral accounts from the early 19th century we find the same drop in activity between about 1800 and 1830 that we see in the sunspot record. In the 17th-and early 18th-century auroral records we find the clear signature of the Maunder minimum, including contemporary accounts from the New World and the Old that express surprise at the dramatic increase in the number of auroral displays that began to occur at the close of the solar Maunder minimum, about 1715. Evidence of earlier periods of prolonged auroral suppression also emerges from the records.

With the fading of the earliest auroral descriptions we have reached the limit of useful, written records that can tell us of the history of the sun. We must search elsewhere.

Tree Rings and Radiocarbon

In the early years of the present century, A.E. Douglass, an Arizona astronomer, endeavored to find a longer and more objective way to recover the sun's past history—using trees. He had noted, as had others, that the width of annual growth rings measured conditions during the year that the ring was formed. In the arid Southwest where Douglass worked, ring width was principally a measure of annual rainfall. The possible tie to the sun, in Douglass's mind, came through a supposed connection between solar activity and the weather. If there was such a tie, there was hope of recovering an earlier history of the sun's activity by measuring the widths of dated rings in trees.

Thus Douglass initiated modern dendrochronology, an indispensable tool of archaeology and climatology. But he never found any clear link between sunspots and tree-ring widths. Yet now we know that there *is* a record of solar history in the trees around us: not in the ring pattern but in the wood itself.

Wood fiber, cellulose, contains radiocarbon or carbon-14, a radioactive isotope. This substance enters the leaf or evergreen needle as carbon dioxide, a vital component of photosynthesis. Then the isotope finds its way into the current year's ring of annual growth. Through chemical analysis we can measure how much radiocarbon each ring contains, and since through dendrochronology we can identify the year of growth of each ring, we can reconstruct how much radiocarbon was in the air long ago, when the tree ring was formed.

The connection with the sun comes through the production of radiocarbon. This form of the chemical element is created high in the Earth's atmosphere when cosmic rays from galactic space collide with nitrogen atoms. Generally, as solar activity increases, the Earth is progressively shielded from galactic cosmic rays, and at such times less radiocarbon is created high in the atmosphere. Naturally, less enters the trees. When the sun actively lags—fewer sunspots—the shield goes down. More galactic cosmic rays reach the Earth, more radiocarbon is formed, and more enters the trees. Thus the radiocarbon content of each tree ring is an inverse measure of solar activity when the ring was formed.

As part of the special efforts made in the 1950s to improve the accuracy of radiocarbon dating in general, the technique of measuring the radiocarbon content of tree rings was perfected. Long-lived trees, like the bristlecone pine, were studied. One of the findings was a pronounced increase in the radiocarbon content of tree rings formed in the late 1600s and early 1700s. It was called the DeVries effect after the Dutch geochemist who first noted it, but now we recognize it as a reflection of the Maunder minimum in sunspot activity. The duration, extent, and nature of the DeVries increase in radiocarbon are exactly what we would expect from the solar conditions that Maunder described. In addition, once corrections were made for solar irregularities, the accuracy of radiocarbon dating was increased, especially for prehistoric times.

Verifying the connection between the Maunder minimum on the solar surface and the DeVries effect in the trees made possible a dramatic breakthrough in the search for solar history. For the first time we could accurately trace the behavior of the sun back before the time of the written word. What is surprising in these first looks into the past is how much the sun has varied, and how frequently periods of prolonged solar depression occur. In the tree-ring radiocarbon data now available, reaching almost 8000 years into the past, we can identify at least a dozen periods when the sun was unlike it is today.

A Further Reach

We would like very much to push the search farther still. If the tree-ring data could be extended to 15,000 years, we could sample what the sun was like at the end of the last

Sun-streaked Mount St. Helens pumps ash into the atmosphere—most recent of great eruptions from volcanoes around the globe. Such clouds may contribute to climatic change. Below: Robert Stuckenrath of the Smithsonian's Radiation Biology Laboratory conducts a radiocarbon study. These tests can determine not only the age of an ancient object but also help document solar and atmospheric fluctuations in the past.

major Ice Age. It would be fascinating to find whether the sun played a role in that great drama of Earth and ice.

Another chemical tool now under study promises even more exciting prospects for revealing solar variation. A French group in Paris and Grenoble are testing the use of the beryllium-10 isotope as another tracer of ancient solar activity. This radioisotope is also formed in the upper atmosphere of the Earth through the action of galactic cosmic rays, and its production rate should also be modulated by surges of solar activity. Unlike radiocarbon, with its useful range of perhaps 20,000 years, beryllium-10 retains its radioactivity longer and can help us look back a million years or more. Beryllium drifts down to the surface of the Earth where it collects in the layered ice deposits of Antarctica and Greenland, and in datable sediments on the bottom of the ocean.

An unexpected record of the really ancient sun may be found in rocks beneath our feet. George Williams, a Melbourne geologist, recently reported a surprising analysis of finely layered stone in southern Australia. The material shows a distinctive repetitive pattern

Master Chart of Solar Cycles

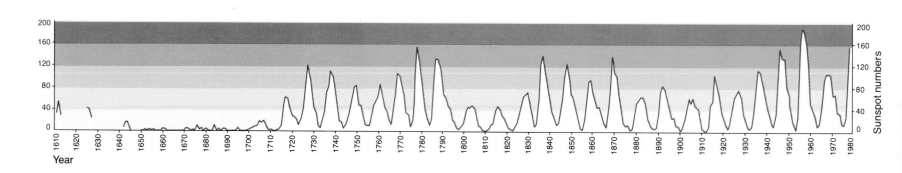

Cyclical desertification across arid monument lands of the United States may be linked to pervasive solar influences. Such climatic episodes helped depopulate advanced cultural sites as at Canyon de Chelly, and New Mexico's Pueblo Del Arroyo, top. Samples of annual growth rings from logs used in pueblo construction may yield clues to the onset of long dry spells. Data comes from both physical examination of the rings and radiocarbon analysis. Results of these investigations, and others from around the world, help refine a master chart of the sun's rhythmic behavior, below left. The short, eleven-year mode has been called the solar heartbeat, and shows the highs and lows of sunspot frequency. Other wavelike variations on a longer time scale suggest themselves and provide clues to solar activity.

in the widths and colors of the bands. Williams demonstrates that the layers are glacial deposits of sediment which formed at the bottom of an ancient sea and that they probably measure annual water run-off. Obvious cycles in the widths and composition of layered bands of sediment bear many resemblances to the modern 11-year sunspot cycle, and the investigator feels they demonstrate a strong control of climate, at the time, by the sun. The 11-year cycles in rock layers persist through the entire 1700-year period of his rock sample. Features like the Maunder minimum also appear. What is most surprising is that the sediments were formed in the Precambrian era, about 670 million years ago.

We cannot be sure that the layers are annual bands or if this is indeed a record of the sun. If it is, however, it is our oldest detailed look at our star. The sun we see mirrored in Australian rock is about 15 percent younger than the one that shines above us today. Yet it looks a lot the same, with its 11-year cycles and frequent sunspot minima of longer duration.

A Long Look

In the longest view, all that we can safely assume is that the sun is well-behaved. We can also be fairly sure that the sun's brightness has been nearly constant for most of Earth's existence. The evidence for this comes from geology and paleontology and our present understanding of the sensitivity of the Earth's climate to changes in total solar radiation. Scientists find liquid water in sedimentary rocks that are 3.7 billion years old and have discovered microfossil life that is 3.4 billion years old. These facts attest that life as we know it existed a very long time ago. This in turn implies that the sun was never so bright that all the oceans boiled nor so dim that ice covered all the Earth. Modern mathematical simulations of climate, coupled with the fossil evidence, refine the estimate and suggest that through the last three or four billion years the brightness of the sun has been constant to within about 10 to 20 percent.

Studies of the soil of the moon and the surfaces of lunar rocks offer similar reassurance. We find that the flow of atomic particles from the sun—protons, electrons, and atomic nuclei—were in the past few billion years roughly what we measure today. These coarse-averaged estimates are confirmed, in a sense, by what we can see in distant, sun-like stars. The really explosive stars are a lot younger, or a lot older, and certainly more massive than our sun. The sky is not full of firecrackers.

Such estimates are comforting but not very specific. We have, in a sense, a report card from school that says our sun has probably not committed any violent crimes, or at least not any that were long enough and severe enough to leave permanent scars on the building. We should like to know a good deal more. For example, the same climate models that put 10-to 20-percent limits on past solar fluctuations also suggest that a five percent drop in solar brightness could bring on a major Ice Age, a one percent drop a Little Ice Age, and a 0.1-percent change a significant alteration of our present climate.

The search for a better and longer history of the sun continues, and with new emphasis. The quest—in books and trees and ocean cores—is more than a search for curiosities. What we learn of changes on the sun tells us something about all the other stars. And the more we know about stellar dynamics—close to home and throughout the universe—the more accurately we can anticipate the full range of what the sun has in store for us, including its influence on the climate of the Earth.

Light and living beings exist in concert on our planet—probably the only one in the solar system to harbor life. Frederic Church, the 19th-century artist whose work appears here, evokes the sun's mysterious life-giving touch in *Cotopaxi*. The painting represents an Andean eruption above the jungle—complete with a llama and its owner.

Observers throughout the ages have shared Church's fascination with the sun's role in the interplay of natural forces. For the Greeks, sunlight epitomized the "fire" in their classical quartet of basic elements—earth, air, fire, and water. They believed that from this metaphysical concoction all living things arose. A respected solar scholar of an earlier day, Georgio Abetti, went so far as to characterize the sun by exclaiming "It's alive." And John Eddy, astronomer and contributor to this book, grows most

Giver of Life

thoughtful as he compares the sun to a throbbing heart. For that matter, so did the Aztec.

Certainly, all the authors in this section of *Fire of Life* affirm the vital benefits of the sun's light. Cyril Ponnamperuma writes of the ancient sun, whose rays he investigates as a source of some early organic molecules on our planet. We visit his laboratory where harsh radiation from a model of the primal sun sparks a chain of chemical evolution.

Robert Carola helps us catch up on a bit of schooling, for scientists have recently revised their idea about how sunlight becomes sugar inside a green leaf. Then, Edward Ayensu of the Smithsonian's Office of Biological Conservation, tells how today's rampant destruction of the tropical forest may trigger damaging climatic changes throughout the world.

John McCosker takes us exploring into the watery realm where we hobnob with fin whales in their ocean-roving quest for krill, then investigate the domestic lives of coral-dwelling denizens. Finally, well-known scientist Sterling Hendricks helps unify our understanding of the sun's work in both winding up and regulating the internal biological clocks and calendars of all creatures, wet and dry.

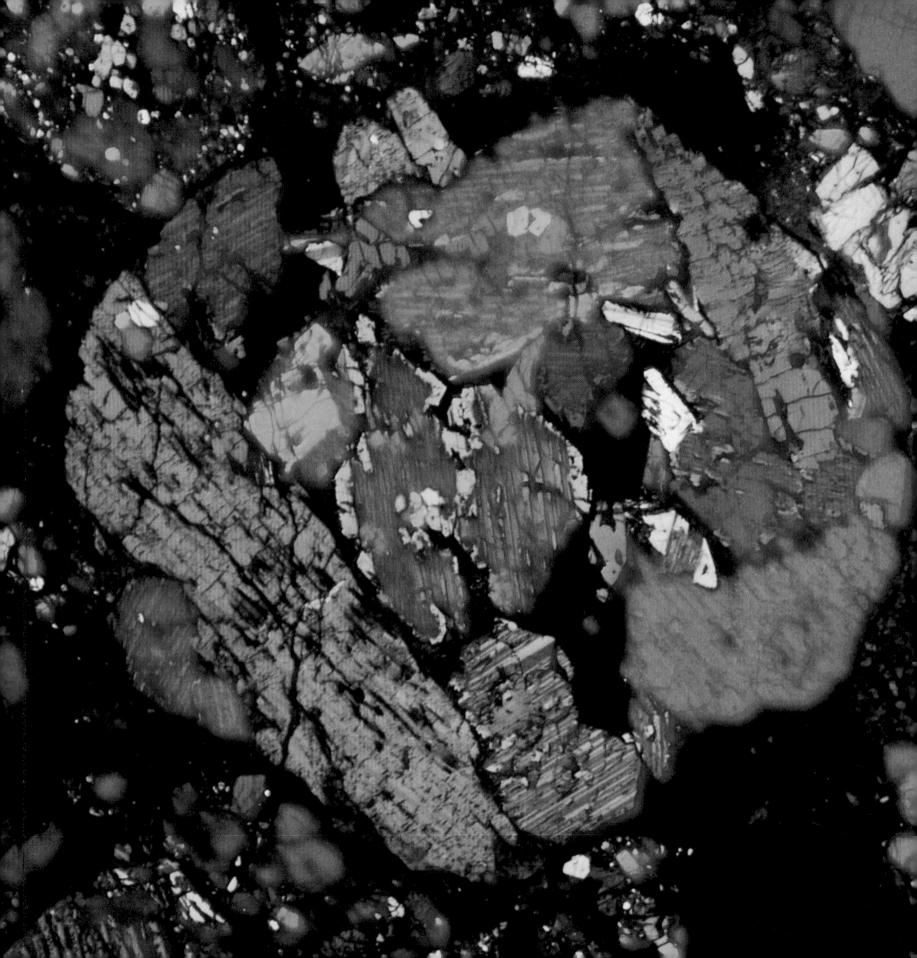

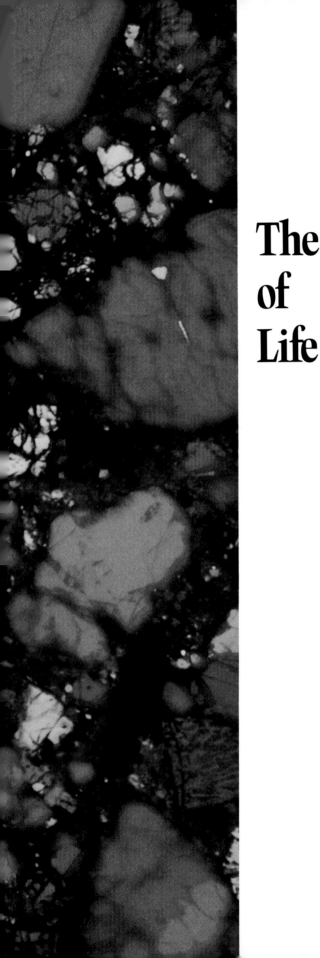

Cyril Ponnamperuma

The Quickening of Life

In a celebrated essay written in 1928, the British biologist J.B.S. Haldane speculated on the early conditions necessary for the emergence of life on Earth:

When ultraviolet light acts on a mixture of water, carbon dioxide, and ammonia, a variety of organic substances are made, including sugars, and apparently some of the materials from which proteins are built up. Before the origin of life they must have accumulated until the primitive oceans reached the consistency of a "hot dilute soup."

This concept of a primordial soup as a necessary preamble to the beginnings of life is at the heart of the modern scientific theory of chemical evolution. The sun, indeed, is the principal actor in this continuing drama.

Prior to Haldane, the Russian biochemist Alexander Ivanovich Oparin in 1924 had alluded to the fact that the complex manifestation and properties characteristic of life must have arisen in the process of the evolution of matter. According to Oparin, the behavior of the organic substances was governed by the properties and arrangement of the atoms which made up molecules. Simple relations between molecules led to more complex associations, climaxing in the appearance of biological order.

The then radical hypotheses of Oparin, Haldane, and others were themselves the products of more than half a century of scientific discovery since Darwin had enunciated his theory of biological evolution. Until the 20th century, living things, particularly those of the so-called lower orders of life, were widely regarded as being the products of spontaneous generation.

"Your serpent of Egypt is bred now of your mud by the operation of your sun; so is your crocodile," exclaims Lepidus to Mark Antony in Act II, Scene VII of *Antony and Cleopatra,*

A photomicrograph, left, of an ultra-thin slice from the Murchison Meteorite, which fell near Murchison, Australia, in 1969. A carbonaceous chondrite, or carbon-rich stony meteorite, the Murchison yielded the first unambiguous evidence of extraterrestrial amino acids, building blocks of proteins in Earthly organisms. Above, ancestors of this simple, seemingly primitive, siphonophore appeared about a half billion years ago, more than three billion years after life began on the Earth.

highlighting the age-old belief that life arose from nonlife through the all-pervading influence of the sun.

One had only to accept the evidence of the senses, thought the ancients: worms from mud, maggots from decaying meat, and mice from old linen. Philosophers of East and West were in agreement on the continuity of matter from the nonliving to the living. In the sacred Hindu scriptures, the beginning of life is attributed to the primary elements, and the ocean is described as the cradle of all life. Aristotle had discoursed in a similar vein in *Metaphysics*, taking as the basis of his observation the apparent transformation of morning dew into the fireflies that swarmed on blades of grass when the sun had set. This teaching was accepted without question by the long line of medieval thinkers who turned to Aristotle as the final authority in matters physical and metaphysical.

The development of a rigorous scientific method called into question the fundamental observations on which the theory of spontaneous generation was erected. An acrimonious controversy raged, and it was not until the middle of the last century that Louis Pasteur, by a series of epoch-making experiments, demonstrated that living organisms did not arise from nonliving material. Although we hold up Pasteur's experiments to beginning students in biology as a triumph of reason over mysticism, we now return to spontaneous generation in a more scientific and precise connotation, namely, chemical evolution.

The raw materials which compose the chemical building blocks of life are the products of stellar evolution, both of our own sun and of earlier generations of stars. The composition of the sun is a blueprint of the distribution of the various elements that make up the solar system. With the exception of inert helium, abundant on the sun, the same is true for the elements that make up the basic constituents of all living organisms: hydrogen, oxygen, nitrogen, and carbon.

The sun and the planets all arose from the same primordial nebula. During the period of the Earth's accretion and formation, most of its original protoplanetary hydrogen-rich atmosphere was probably driven off by the sun. As the core of the Earth warmed up, the volcanoes that blistered its newly formed crust must have belched out trapped gases which then became the "primitive" atmosphere. A large amount of cosmic debris falling onto the Earth's surface, or attracted by its gravitational forces as it orbited the sun, further enriched this primitive atmosphere.

The water vapor escaping from the volcanic vents of the Earth, or raining down from interplanetary space, condensed and accumulated to give rise to the oceans. The scene was now set for the rays of the sun to play upon atmosphere and ocean and give rise to the "hot dilute soup" which Haldane so ingeniously described.

Though there is still some uncertainty about the nature of the primitive atmosphere, it is generally believed that it was very different from that existing today. A number of factors lead us to believe that the carbon, nitrogen, and oxygen enriched by hydrogen resulted in an atmospheric mixture of methane, ammonia, and water—a noxious concoction by present standards but an ideal composition for interaction with the penetrating ultraviolet rays of the sun.

In the solar system, the occurrence of free oxygen is a characteristic unique to the Earth. This free oxygen, which makes up almost a fifth of our present atmosphere, is very largely due to the photosynthetic process carried on by plants. The evolutionary development of photosynthesis, by which an organism was able to capture a photon of energy from the sun and use it to convert components of the atmosphere into sustenance for itself, could be considered a result of dire and drastic conditions produced on the planet by the action of the sun's rays on water vapor in the upper atmosphere.

Some four billion years ago, volcanoes belched forth gases— mostly methane, ammonia, and water vapor— hitherto trapped in the Earth's newly formed, still barren crust. From this primitive atmosphere, some water vapor condensed and fell to fill the ocean basins. The seas, enriched by interaction with the atmosphere and by minerals leached from the land, were bombarded by ultraviolet radiation from the sun. The result: the seas became a "hot, dilute soup" of organic molecules. These were the precursors of the self-replicating molecules, the first living things.

When that water vapor was split into its components, hydrogen and oxygen, by the action of the sun's powerful rays, or photodissociated, the lighter hydrogen must have escaped into space. The oxygen atoms, however, recombined as ozone, which is opaque to ultraviolet radiation from the sun. Before the formation of the ozone layer, these ultraviolet rays had penetrated the atmosphere and played on the surface of the seas, producing the rich supply of organic molecules upon which the earliest living organisms must have "fed." Called heterotrophs, organisms that obtain their food from organic substances, these living entities were very different from their counterparts of today, depending for sustenance as they did entirely on the organic molecules created by the interaction of ultraviolet radiation with the primordial soup of the seas.

With the buildup of the ozone layer and the shielding of the Earth's surface from most ultraviolet radiation, a wholesale disappearance of life must have taken place. The heterotrophic organisms no longer had a constantly renewed supply of organic molecules to feed upon. If these pristine organisms were alive today, they

As the ribbon of time and life indicates, life on Earth is a Johnny-come-lately, cosmically speaking. Perhaps as much as three fourths of all of time since the Big Bang passed before the sun and planets formed. After that, however, life appeared on Earth very quickly. Since then, life and planet have evolved together in a relationship apparently unique in our cosmic neighborhood.

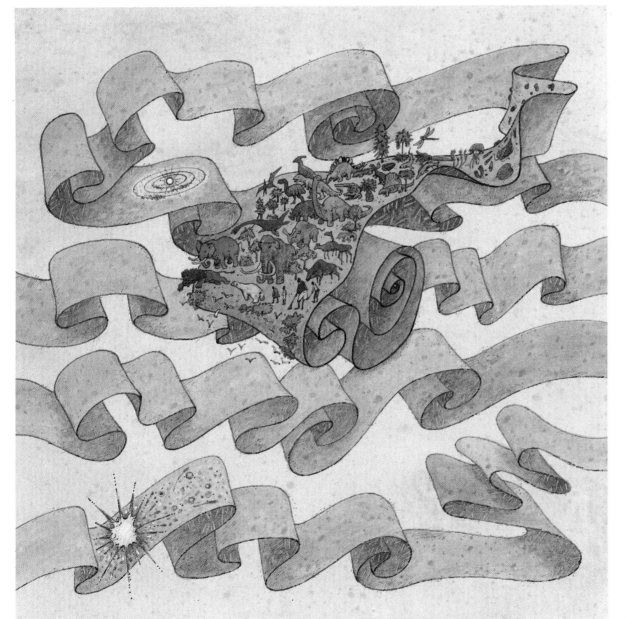

The hourglass-shaped diagram above represents the evolution of the atmosphere and the near extinction of life early in the Earth's history. The first life forms "fed" on organic molecules created by the interaction of ultraviolet radiation with the seas and atmosphere (the lower, reddish part of the diagram). As the ozone layer (the narrow neck of the diagram) began to form, shielding the Earth from ultraviolet, the "food" supply was cut off, and many life forms died out. Fortunately, some evolved the ability to photosynthesize. In the process, they gave off free oxygen, gradually changing the primitive atmosphere, rich in hydrogen compounds, into the modern, oxygen-rich atmosphere (upper, green part of the diagram).

would be clamoring "Down with the ozone layer!" It is ironic that this layer of ozone, the very cause of the destruction of early life, is today the most powerful safeguard for contemporary organisms.

However, the evolution of the life forms capable of photosynthesis saved the day for life on Earth. And these same organisms were the reason for the dramatic change from an oxygen-poor to an oxygen-rich atmosphere. The ozone shield, while interrupting the passage of the ultraviolet light from the sun, was not capable of interfering with the rest of the sun's energy—the longer wavelengths of light. The photosynthetic organisms were able to absorb this light and make their own food, thereby becoming the first "autotrophs," and, in the process, releasing molecular oxygen into the atmosphere.

As we have already seen, a great deal of solar energy is blocked by the ozone layer, but in the very early days of the Earth's existence much of this energy reached the Earth's surface. Therefore, in laboratory studies simulating conditions on the prebiotic Earth, extensive use has been made of this form of energy, particularly in the energetic ultraviolet wavelengths.

During the last two decades, a number of experiments have been performed in laboratories around the world to recreate the effect of ultraviolet light on the Earth's primitive atmosphere before the appearance of life. In a typical experiment, the process was re-enacted in a dumbbell shaped flask in which the upper chamber represented the atmosphere and the lower one the ocean. A mixture of methane, ammonia, and water vapor—the ingredients of the primordial atmosphere—filled the upper chamber, while the lower glass vessel contained water. A beam of ultraviolet light was played on the upper part of the flask representing the atmosphere. A variety of organic molecules collected in the lower flask, including the components of nucleic acids and proteins which are at the basis of all life.

Photochemists, or those who study the chemistry associated with light energy, have devised a number of lamps which act as solar simulators. Some of these pinpoint light at a particular wavelength while others reproduce the entire energy spectrum of the sun. A particularly versatile lamp is one which was designed by NASA during the Apollo program. In this particular device, argon gas maintained at a

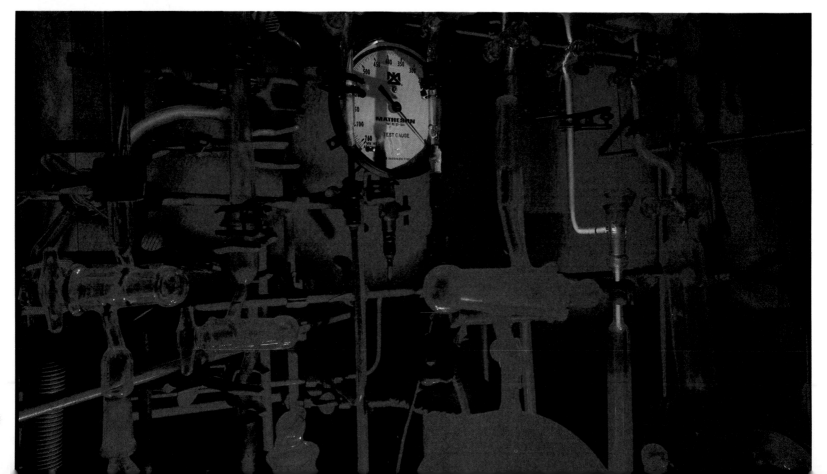

high pressure in a closed chamber was energized with a massive input of electrical energy. The beam of ultraviolet light thus generated was 10,000 times more concentrated than solar radiation reaching the Earth in those wavelengths. When directed into a vessel containing gases believed to represent those comprising the Earth's primitive atmosphere, the highly energetic rays duplicated in a few hours the effects of thousands of hours of sunlight falling on the ancient Earth. It was like an Aladdin's lamp, the answer to a photochemist's prayer.

Many precursors of the building blocks of life have been detected in the heads and tails of comets. Perhaps the single most important compound among these is hydrogen cyanide. This molecule results from a combination of the atoms hydrogen, carbon, and nitrogen and is written HCN. In today's atmosphere of free oxygen, hydrogen cyanide is a powerful poison that interferes with the respiratory mechanism and is a speedy agent of death. However, in prebiotic times it was an essential ingredient of the mixture leading to life.

When a very weak solution of hydrogen cyanide is exposed to ultraviolet light, a large number of complex organic molecules are formed. Among these compounds are adenine and guanine, which are components of the nucleic acids, and several amino acids which are the building blocks of protein molecules.

Not only have the constituents of the proteins and nucleic acids been made in the laboratory by simulating the action of the sun, but the next step in the process, the linking up of these units into chainlike molecules, has been achieved. It is thus apparent from exhaustive laboratory studies that solar energy must have played a decisive role in the first steps that led to life on this planet.

Our search for the origins of life and the possible role played by the sun takes us back across the eons to detect the scanty evidence of the earliest living organisms on Earth. Is it even possible that sequestered in some cranny of an ancient sediment could be found vestigial evidence of the primordial molecules generated by solar energy?

As the Earth's surface has been continually built up and torn down over the eons, such tiny, primordial fossils are not easy to find. In fact, because no skeletons or other hard fossil parts had been found to pre-exist 600 million years ago, generations of scientists assumed

At the University of Maryland's Laboratory of Chemical Evolution, the author and colleagues reproduce the primitive Earth's environment with a variety of experiments using lamps to simulate solar radiation. Below, in ultraviolet, this broad spectrum "Aladdin's Lamp" shines with an intensity 10,000 times that of the sun into flasks containing gas mixtures thought to resemble those of the primitive atmosphere. Narrow spectrum solar simulators, opposite, are employed to explore the behavior of particular molecules as they absorb energy. Below left, rarefied gases glow eerily in a gas transfer apparatus used to fill flasks with mixtures like those of the ancient atmosphere.

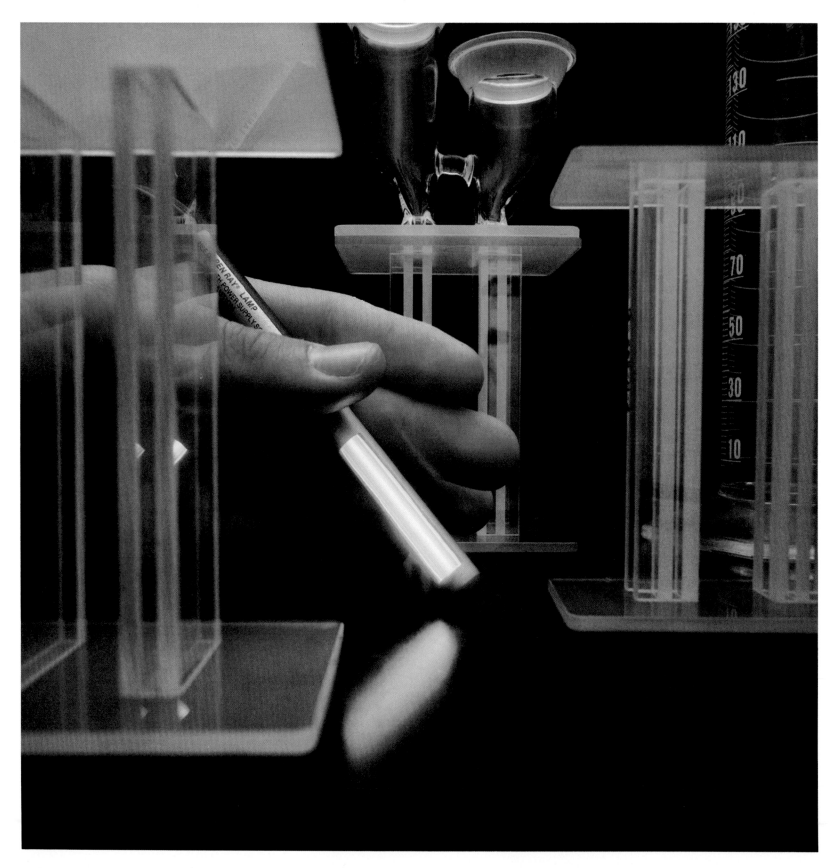

that no life existed before that time. However, the monumental work of Elso Barghoorn of Harvard University and his students has unfolded for us an evolving pattern of life from almost the very dawn of terrestrial time.

Around one billion years ago, the microfossil-rich Bitter Springs Formation of the Northern Territory of Australia was formed. Reaching back another billion years, we have a remarkable example of a diverse bacterial life in the Gunflint chert of Ontario, rocks which occur in the northeastern part of the United States and southeastern Canada, some protruding into Lake Ontario. It was in the study of these specimens that Professor Barghoorn first turned up the microfossils of the Precambrian era.

Almost a billion years older than *these* sedimentary rocks of Ontario, the Soudan formation, another outcrop in the Montana and Minnesota region of North America, allows us a glimpse of the bacterial population in existence at that time.

Of paramount importance are the rocks from the Swaziland outcrops of southern Africa. The Fig Tree shale, dated at 3.1 billion years, and the Onverwacht at around 3.5 billion have both yielded, in the hands of Professor Barghoorn and his associates, evidence of early life on

Gas mixtures can be irradiated with an ultra-narrow spectrum solar simulator such as the 1849-angstrom wavelength mercury lamp on the opposite page. These hand-held sources are also used to analyze irradiated gases. Below left, a sample of the Earth's oldest known rock, from a 3.8 billion-year-old sedimentary formation in Isua, Greenland, in which the author has found nearly conclusive evidence of carbonaceous material of biological origin. The green colonial algae, Hydrodictyaceae, below, are relatives of the blue-green algae, oldest known photosynthetic forms on Earth. Some may have lived 3.5 billion years ago.

Earth. All these organisms appear to be morphologically similar to blue-green algae, present-day microscopic photosynthetic organisms using solar energy.

More recently has come fossil evidence of bacterial colonies from a find in Western Australia. Stromatolites, or fossil colonies, are produced when filamentous photosynthesizing bacteria push their way through an overlying layer of debris in search of the sun and thus build up layers piled one above the other. Before the Australian discovery, dated at 3.5 billion years, the most celebrated of these finds was the Bulawayan stromatolite from the southern African formation, dated at 2.7 billion years ago.

Scientists investigating the barren, rocky formation along the ice sheet of Greenland have unearthed a spectacular sequence of sediments at Isua, northeast of Godthab. These deposits are truly sedimentary, having been deposited in the presence of water, and they have been dated at 3.8 billion years. These are indeed the oldest known sediments on Earth. The Eskimo word *Isua* aptly translates, "the farthest you can go." Careful analyses of these rocks in the laboratory have revealed evidence of possible life at this early period—indeed, life that appears to have been photosynthetic. As far back as we can go with fossil evidence, life appears to have been prevalent—and largely dependent on the sun.

As we look at the night sky and see countless stars twinkling, we cannot help thinking that many of the stars are like our sun. Since our sun is the mainstay of life upon this Earth, there must be life on other planets around other suns.

Such a conclusion, which astronomers have reached by the rigorous analysis of scientific data, was prophetically foretold by Giordano Bruno in the 16th century:

Sky, universe, all-embracing ether, and immeasurable space alive with movement—all these are of one nature. In space there are countless constellations, suns, and planets; we see only the suns because they give light; the planets remain invisible, for they are small and dark. There are also numberless earths circling around their suns, no worse and no less inhabited than this globe of ours. For no reasonable mind can assume that heavenly bodies, which may be far more magnificent than ours, would not bear upon them creatures similar or even superior to those upon our human earth.

Robert Carola

Sunlight to Sugar

No form of life can exist without a source of nourishment. For life on Earth, that source is the sun. But sun rays have no food value in themselves: plants alone have the ability to trap light energy from the sun and convert it into the chemical energy which they transform into food. There is simply nothing else on Earth that even vaguely resembles this capacity of plants to hold and fix the power of sunlight into usable forms. Without it only the most primitive living things could exist. This chemical reaction—perhaps the most important in nature—is photosynthesis. And giving it a scientific name and analyzing it in no way detracts from its fundamental wonder, for all life totally depends on this complex yet common phenomenon.

We still do not fully understand how plants accomplish their basic conversion of light into food, but since the discovery of photosynthesis by Joseph Priestley about 200 years ago, our knowledge has grown a great deal. And we have made rapid gains since 1930.

Plants can store enough food energy through photosynthesis to supply not only themselves, but animals as well. And those animals that do not eat plants must eat animals that do. The Bible says that *all flesh* is *grass*.

Besides getting much of their food from plants, animals also obtain oxygen—the air they breathe—as a by-product of photosynthesis. So without light from the sun, and photosynthesis here on Earth, plants and animals could not exist.

Photosynthesis becomes more understandable if we analyze the word. *Photo* means light, and *synthesis* means the process of building up a complex substance through the union of simpler substances. The key is a green pigment in plants called chlorophyll. Generally, light energy striking the chlorophyll in a plant is

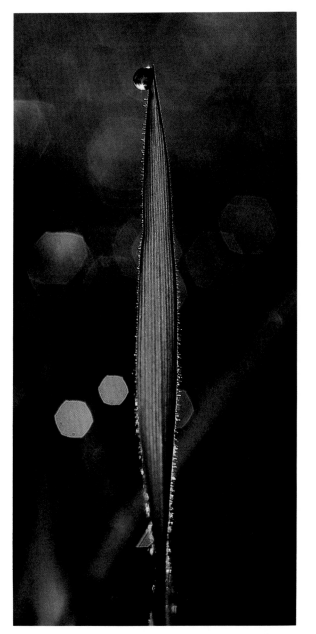

Sunlight reveals the complex vascular structure of a cottonwood leaf, opposite. Channels conduct water from the roots to areas with chlorophyll and circulate sugars produced by the green catalyst. Grass blade exudes a drop of excess water in a process called guttation. In leaf or stem, chlorophyll-bearing cells are specialized factories for converting sunlight to food substances.

converted into chemical energy. The plant takes simple raw materials, water and carbon dioxide, and transforms them into energy-laden molecules such as sugar in the form of glucose. Oxygen and water form the major by-products.

The chemical process of photosynthesis may be outlined in this way: six molecules of carbon dioxide combine with 12 molecules of water in the presence of light energy and chlorophyll to form one molecule of glucose, six molecules of oxygen, and regenerate six molecules of water. In their normal states, carbon dioxide and water are useless to plants because they do not oxi-

dize, or burn, and are unsuitable as food. Glucose, however, is an excellent food and oxidizes readily to release usable energy.

Scientists have known for more than 50 years that at least two steps take place during photosynthesis, one requiring light and one independent of light. This distinction is well worth an extra moment of concentration as it has to do with a most important discovery about photosynthesis: water molecules composed of hydrogen and oxygen were being split by light to provide oxygen. A great deal of energy is required to split water molecules, and scientists

Approximate Magnifications

X150   X600   X 6,800

In an artist's conception, seven progressive magnifications reveal the increasingly minute machinery of photosynthesis.

Above left, leaf cross section shows the mesophyll tissue, with a layer of compact palisade cells, and a layer of irregularly shaped spongy cells below them. Both kinds of cells contain numerous chloroplasts, tiny capsules of chlorophyll, a green pigment whose catalytic reaction to light initiates the process of photosynthesis. Above, a higher magnification of several palisade cells spotlights the chloroplasts.

Electron photomicrographs further reveal the chloroplast's complex internal structure. Visible here are the membranes—called lamellae—that run parallel to each other and extend the length of the chloroplast, and the stacks of flattened, coin-shaped disks called grana. These appear to provide the inert physical support—substrate—that holds the dispersed and highly active chlorophyll molecules.

soon concluded that not only did light provide the energy for the production of glucose, it also powered the separation of oxygen from water. The photosynthetic reactions that take place in the light—the trapping of light by chlorophyll and the splitting of water molecules—are called the light reactions. Carbon dioxide is required in the dark reactions in which sugar, oxygen, and water are produced.

Questions of terminology may create some problems. But the confusion is likely to vanish if we keep these points in mind: as we might expect, the light reactions take place only in the light; the dark reactions, however, do not necessarily take place in the dark or at night. "Dark" simply means that such reactions may proceed without light. Basically, the light and dark reactions are consecutive steps in the overall process of photosynthesis. It starts with the trapping of solar energy in the light reactions, and ends with the production of sugar in the dark reactions. And both go on simultaneously, photosynthesis being a continuous process under suitable environmental conditions.

The light and dark reactions are separated here only to stress that the light-absorbing

X35,000

X120,000

X 650,000

X3,500,000

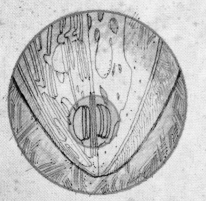

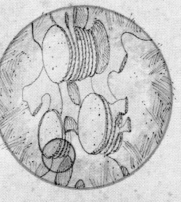

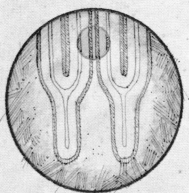

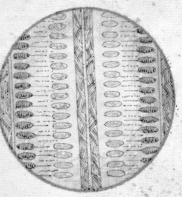

A section through two grana highlights the neatly arrayed stacks and the lamellae, or membranes, of the grana. The regular arrangement of elements presumably helps to orient chlorophyll molecules to sunlight. Molecules themselves contain magnesium, an active metallic element that may encourage the catalytic energy transfer.

Still closer, a three-dimensional representation pinpoints the individual membrane-bound disks of several grana. Similar to other cell membranes in some respects, these thin, specialized grana lamellae contain the chlorophyll itself and other associated organic and inorganic compounds necessary for the "light" reactions of photosynthesis.

Two disks of a granum reveal layers of chlorophyll sandwiched between outer layers of protein. The clear substance that surrounds the grana in the chloroplast is known as the stroma, and is the site of the so-called "dark" reactions of photosynthesis. These reactions—termed dark because they do not require light in order to proceed—use chemical energy to produce complex molecules of sugar and starch.

Broken down to its major chemical constituents, chlorophyll molecules line up alternately with lipid and carotene molecules within an individual lamella of the granum. The boundary lines represent protein. Carotene, an accessory pigment, absorbs certain wavelengths of light and thus transfers additional solar energy to the chlorophyll.

Ancient allies— dragonfly and horsetail plant —survive from the Carboniferous, an era of great coal formation. Many closely associated plants and animals evolved in tandem. The horsetail's hollow, jointed stem—part of a vascular system—represents one of the primitive adaptations that enabled plants to emerge from the sea onto dry land.

130

activity is a photochemical event, while the "dark" event is a standard chemical transformation. The light reactions convert light energy to chemical energy; the dark reactions use chemical energy to produce complex organic molecules.

The entire process of the light reactions occurs in the fraction of a second that it takes sunlight to activate chlorophyll. Electrons in the chlorophyll are raised to the higher energy level needed for them to escape from the chlorophyll molecule. These energy-carrying electrons ultimately accumulate in an organic compound, NADP, which is an electron-accepting coenzyme in chlorophyll. NADP (nicotinamide adenosine dinucleotide phosphate) accepts not only electrons but also the hydrogen atoms that are released from water when the chlorophyll is activated or "energized."

In the electron-accepting process, NADP is converted to $NADPH_2$, which carries its load of hydrogen and electrons to the dark photosynthetic reactions. Here, hydrogen helps in the formation of carbohydrates.

Meanwhile, the electrons from chlorophyll have left a "gap" in the chlorophyll molecule, and that gap must be refilled before photosynthesis can continue. The electrons lost from the chlorophyll are replaced by electrons from new water molecules that are split into protons, oxygen, and electrons. These electrons fill the gaps left in the "energized" chlorophyll, and in this way they help keep the process operating. NADP may be reduced over and over to $NADPH_2$ as it accepts and then liberates electrons and protons. The oxygen atoms out of water constitute the familiar oxygen molecule O_2, which diffuses into the atmosphere.

At the same time, another important action is going on. As the electrons pass along their chain of carriers, they become less energetic at each step, although not all their energy is lost to the plant. The energy that is not used in splitting water molecules goes on to help make an important energy carrier, ATP (adenosine triphosphate). It is created from chemicals in the plant's chlorophyll-containing structure: one of these being ADP (adenosine diphosphate), a substance from which ATP—the driver molecule—is made.

The dark reactions of photosynthesis take place in the chloroplasts. There, complex molecules of sugars and starches are synthesized from simple carbon dioxide with active hydrogen from $NADPH_2$. The round of dark reactions is named the Calvin cycle, after Melvin Calvin, who outlined the separate steps during the early 1950s, receiving the 1961 Nobel Prize in chemistry for his work.

Some desert plants and some grasses, such as sugar cane and crab grass, do not use the Calvin cycle exclusively. Theirs is a variant termed the 4-carbon cycle, and it is most effective in tropical or desert conditions with intense light, high temperature, and limited carbon dioxide. It has been known for a long time that a sugar cane field is one of the most productive energy traps in all of nature, and part of the explanation is that sugar cane uses 4-carbon photosynthesis. Maize, another superproducer, also capitalizes on this variant cycle.

Light is essential for the completion of photosynthesis, but not all light is equally effective. The chlorophyll pigments absorb some radiation along the entire visible spectrum, but more at the long wavelength, or red, end. In fact, plants appear green because they absorb red light and reflect the green portion of the spectrum.

The intensity of light, which is the amount of energy delivered on a given area per unit of time, is as important as the color, or "quality,"

Spore-like bodies and filaments from the Gunflint chert offer fossil evidence of the most ancient green plants known today, unicellular algae that lived 2000 million years ago.

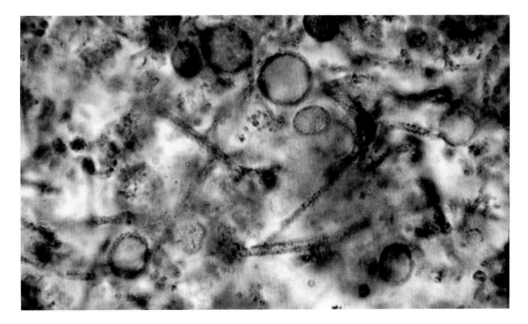

Specialized leaves capture blue-light wavelengths, enabling a peacock fern of the dim jungle floor to absorb enough light to make its food. Like some algae in ocean depths, this Asian land plant derives its blue color from organic structures rather than from a pigment. Living sundial, a sunflower from a medieval book follows the solar light across the sky.

ΩΡΟΣΚΟΠΙΟΝ ΕΛΙΟΤΡΟΠΙΚΟΝ.

Artis et Naturæ Coniugium.

of the light. If that intensity is too low, photosynthesis stops. At the other extreme, the light can reach such high intensity that the plant becomes saturated. Many plants, especially those adapted to the low light intensities of deep water or tropical rain forests, can be harmed by too much light, and their chlorophyll destroyed.

Chlorophyll's special feature is its ability to be "excited" by certain wavelengths of light. During such excitement, which lasts less than one billionth of a second, pairs of electrons are pushed farther away from their nucleus and are thus raised to a higher, unstable energy level. In some cases the excited chlorophyll electrons merely drop back to their original energy level and produce fluorescence or heat, accomplishing nothing of chemical value. If a suitable electron-accepting molecule is available, however, an excited chlorophyll electron can pass from the original chlorophyll molecule to a suitable chemical acceptor, and photosynthesis will have begun.

Besides chlorophyll, a plant needs carbon dioxide and water for photosynthesis. Although there are only three or four molecules of carbon dioxide in every 10,000 molecules of gases in the Earth's atmosphere, that is enough for plants to carry on the synthesis of sugar. In fact, a modest level of CO_2 is optimal. If the concentration goes too low, photosynthesis ceases. Too much carbon dioxide can poison plants. Plants could actually use much more than they usually receive, even a hundred times more, a concentration seldom found in nature. Were it economical to do so, crops could be "fertilized" by adding carbon dioxide to the air around leaves during daylight. In fact, this readily produced gas is sometimes added to greenhouses for orchids and other prime specimens.

Water is used by plants in many ways, but it has a special value in photosynthesis. Water furnishes hydrogen, one of the two raw materials from which sugar is synthesized, the other of course being carbon dioxide. In an active chloroplast, water is split into its chemical building blocks—hydrogen, two electrons, and an oxygen atom. The hydrogen and electrons become incorporated into an energy-rich sugar molecule, and the oxygen atom combines with another oxygen atom to make molecular oxygen, which escapes into the atmosphere. This is the oxygen animals and people breathe.

Now that we've examined each element of photosynthesis in turn, we can view this key chemical process as a whole, from start to finish, as it normally occurs in the plant. To sum up: water molecules brought into chloroplasts from the plant's roots come in contact with chlorophyll. Sunlight passing into the leaf strikes the chlorophyll, giving it the energy to break the water molecules apart and to supply energetic electrons for photosynthetic activities. At this stage the oxygen from the split water molecule diffuses out of the leaf and the hydrogen stays in the chloroplasts—combined with a driver molecule that has absorbed the excess energy. The energy-laden driver molecule causes the hydrogen to combine with simple carbon compounds already built up in the chloroplast from carbon dioxide in the air. With this last reaction, the change from light energy to chemical energy is complete—a complex, stable, energy-rich carbohydrate has been formed. Energy is the foundation upon which all life is built; the food chain ultimately reaching back to the sun. But Earth's contribution is considerable, too, for chlorophyll is sunshine's potent partner in creating an entire living world.

From a tulip's petal pink to bold sunflower yellow, colors represent chemical processes at work in plants, often in response to light. One blue pigment tests the light, effectively forecasts the weather, then regulates plant metabolism to make the most of sun or rain. Attuned to light, sunflower eyes turn in unison to follow their star.

Edward S. Ayensu

The Terrestrial Greenhouse

Top: curious cat peers around the side of an early terrarium, invention of a London physician in 1829. Nathaniel Bagshaw Ward soon realized that his little inner-space capsules were microcosms of the entire living planet. Opposite: viewed from a spacecraft, the Earth appears to have its own glass roof, the atmosphere. As sunlight goes in through the window of air, it turns to heat which is held for a while and thus powers the global weather system. Carbon dioxide tends to retard loss of heat into space—the more CO_2 near the earth, the more heat retained, right.

For a moment imagine yourself the inhabitant of a terrarium, or Wardian case. At first glance your growth chamber may appear too simple to evoke much interest—merely a miniature garden. But upon closer inspection you would probably observe that your little case holds a balanced community of life, indeed a microcosm of a relatively complex biosphere powered by the sun. Without inspection at all you would know that it is warm and growing warmer as the day wears on, especially if it is left in the bright sun too long. And—like a closed automobile—it could become fatally hot under certain circumstances.

As with a glass-walled terrarium, literally all of the planet's energy comes from the sun. As this shortwave radiation—raw sunshine—heads earthward, it passes through the ozone layer which absorbs most of the ultraviolet rays. Atmospheric water vapor absorbs the infrared rays, and the sun's energy is transformed into long-wave energy, or heat, and eventually radiated back into space. Since carbon dioxide at the Earth's surface absorbs long-wave radiation, the more carbon dioxide in the air the more heat the planetary atmosphere can retain and the warmer our living space can become. This phenomenon is popularly known as the "greenhouse effect."

During the past 100 years or so, mankind's progressive use of fossil fuels appears to have raised the concentration of carbon dioxide in the atmosphere by up to 10 percent. It is now obvious that so much carbon dioxide at the Earth's surface can act as a one-way filter, permitting the transmission of incoming energy from the sun, but preventing its escape back into space. Thus, a warming trend may already be underway on Earth, creating a hot, moist atmosphere like the one you experience in a glass-covered greenhouse. And who could

predict the final toll if such a process upsets the natural cooling of the planet. Mix this with man's apparent destruction of the Earth's protective ozone layer and you have a recipe for global disaster.

All the Earth's biomes—great ecological realms such as desert and tundra—have evolved and continue to maintain themselves by the power of the sun. Their survival, however, depends upon the amount of sunlight and heat that each biome can bear. The plant and animal communities within deserts, for example, have many physiological and behavioral adaptations that allow them to thrive in so harsh an environment. The extent of vegetation cover in deserts and other arid regions depends upon the plant life's ability to tolerate drought. Such plants are generally known as xerophytes (from two Greek words, *xeros* meaning dry and *phyto* meaning plant).

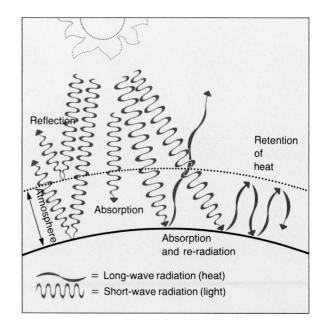

Reflection

Atmosphere

Absorption

Absorption and re-radiation

Retention of heat

⌇⌇⌇ = Long-wave radiation (heat)

ᠱᠱᠱ = Short-wave radiation (light)

Plants adapt to varied climes. At bottom, a South American bromeliad holds rainwater. Virginia wildflower sinks its roots into soil enriched by heavy leaf fall. Tropical forests shed leaves a few at a time and the nourishment returns almost overnight to the trees. Right: Saguaro cactus conserves moisture within its barrel-like body.

Yet there are several plant species growing in desert areas that do not experience direct confrontation with drought—the desert annuals, or ephemerals. During the dry season these plants exist in the form of seeds that can survive high temperatures without losing their ability to germinate. When enough moisture falls on their sandy or rocky soils these seeds germinate and the plants grow very rapidly, so that they complete a life cycle from seed to seed in 30 days.

Some of the ephemerals are so small that they end up producing only one seed. These "belly plants," so-called because one must lie on his belly to study them, are known to possess water-soluble inhibitors in their seed coats that leach out, permitting the seed to germinate only after a heavy rainfall. Other desert plants develop very deep rooting systems to reach the water table and thus manage to avoid the perennial droughts that occur in some desert areas. Alfalfa and mesquite have been found to possess root systems penetrating the soil as far as 129 feet and 175 feet respectively.

Other structural adaptations help plants to escape droughts. These include the dense, hairy coverings which help retain moisture, and the development of small, specialized leaves that help to keep the plant's temperature close to air temperature. Perhaps the two most efficient survival devices adopted by plants in harsh desert environments are observed in cacti, the African desert euphorbias, the century plant *(Agave)*, and other successful succulent arid and semi-arid plants. First, all these plants have developed a special kind of tissue that consists of large thin-walled cells capable of storing water. The outer group of cells is covered with a dense epidermis and a water-impervious, waxy cuticle. Secondly, these plants enjoy what is known as succulent metabolism, opening their stomates at night when the atmosphere is cool and humid and closing them during the day when transpiration stresses are extremely high. In the evening coolness, carbon dioxide is transformed within the plant into organic acids. During the day these are converted by photosynthetic processes to other metabolic products, such as carbohydrates.

Arctic and mountain tundra offer another example of a biologically extreme environment. In the alpine tundra, light intensity is almost always very low. This results from cloudy conditions and the slant of the sunlight especially during the short growing season. The atmosphere is almost always cool, and winds cause much drying of plant tissue. Alpine plants are physiologically adapted to achieve maximum photosynthesis rates at relatively low temperatures—about 15°C. Some alpines such as the sorrels and the non-flowering lichens can photosynthesize even at temperatures several degrees below freezing.

The tropical rain forest represents another of the Earth's major biomes, and it enjoys a strong and almost constant supply of sunlight. Here the simple rule is wherever sunlight falls,

there is chlorophyll to trap it. By photosynthesis the leaves of the forest canopy convert to chemical energy almost as much solar energy as they receive. Using carbon dioxide and water, the green plants transform sunlight into sugar and release oxygen. Because temperatures are high and rainfall is usually abundant, vegetation proliferates. Productivity is extremely high.

Tropical rain forests, unlike temperate forests, consist of five distinguishable layers. Of these, the top three layers consist of trees. Shrubs constitute the next layer, and in the fifth layer herbs and ferns predominate. The topmost trees, forming the emergent discontinuous layer, reach heights of 130 feet or more. The second layer, the canopy, is composed of trees about 90 to 100 feet high, their tops touching in continuous formation. The limited light supply below the canopy allows little vegetation in the ground layers of equatorial forests. Shorter trees of 30 to 50 feet in height struggle for light in

Rose Angel, of the Royal Botanic Gardens in Kew, England, displays one of the few remaining Wardian cases from the early 1800s. Miniature greenhouses like these helped spread tea, rubber, quinine, and other valuable plants to places where they could be commercially grown. Household terrariums started the Victorian "fern craze" and helped spawn important horticultural research. Today, research into jungle ecology may be the best way to save the tropical biome, endangered by human activity.

Road work has uncovered the typically thin topsoil of the jungle to fast erosion by rainfall, above. Live stilts and flying buttresses shore up a shallow rooted African kapok tree.

the shady middle layer. These scattered trees generally have crowns that are narrow in proportion to their height. This adaptation enables them to take maximum advantage of the little sunlight that penetrates the canopy. Dwarf trees reach a height of about 15 feet in the shrub layer.

The layer at the ground nurtures mostly small growth. Many ferns appear, and their leaves intercept a high proportion of the minimal light that reaches them. In many cases the only plants that can survive here are saprophytes, plants which lack chlorophyll and live on the decayed remains of other plants. Parasitic plants here gain their food supply from the roots of other plants. The most spectacular of these is *Rafflesia arnoldi,* which produces the largest flower in the world. It blooms in near darkness.

The lack of sunlight at the forest floor generally impedes the successful growth of seeds dropped by the large trees. As an adaptive response, most of the trees produce many fruits and seeds, and these contain large amounts of nourishment to sustain germination. The forest floor is generally littered with fruits and seeds at different levels of development. Such seedlings, however, do not grow as fast as they might, owing to the gloom. When a tree with a substantial canopy dies or is removed, the open space thus created allows the sun's rays to reach the forest floor, permitting the seedlings in that particular spot to race for the skies. Naturally, not all the saplings can fill the niche. Usually only one or two manage to make it to the top, and as soon as the canopy recloses, the futures of the rest of the seedlings are sealed.

All five layers of flora have faunal elements that are adapted to their particular vertical habitats. The emergent layer often attracts butterflies and birds of prey. The canopy layer nurtures leaf-and-fruit-eaters: parrots, gibbons, flying squirrels, tree boas, lemurs, and howler monkeys abound. The middle layer accommodates leopards, monkeys, civet cats, flying

insects, some bats, and more snakes. The shrub layer harbors nocturnal animals such as owls and kinkajous, while tapirs, brocket deer, flightless curassows, and many other jungle animals browse at the herb layer.

Tropical rain forests represent the most diverse of all of the world's ecosystems from the standpoint of animal and plant populations. For example, in the rain forest in Malaysia a hectare (2.5 acres) may contain between 100 and 150 tree species. In some areas of Brazil's Amazon Basin as many as 300 tree species may be encountered within a hectare. There are thousands of epiphytic plants such as orchids and bromeliads that grow on the trees and take advantage of the sunlight available in the forest canopy.

Several reasons for this great diversity of species have been put forward by some of the world's outstanding tropical biologists, but no single explanation seems to suffice. In any case, all explanations must take into account the various environments in which tropical forests have developed over the years. Some biologists argue that the diversity of tropical forests is due to the constant warm and moist conditions which occur over large areas. Such environmental uniformity, however, might logically be expected to produce vegetation of monotonous sameness. And nothing could be further from the truth.

Perhaps the single most important reason for the great diversity of the tropical forests is the amount of time in which such forests have had to develop. Even though these tropical regions have experienced small-scale paleoclimatic changes over the years, for the most part they have been spared the widespread changes that occurred in the north temperate zones during repeated Ice Ages. Trees may well require 100,000 years or more to develop such a state of homeostasis as is encountered in the mixed forests of the tropics.

Equally significant for diversity is the tropical forests' ability to accept "immigrants" from neighboring biomes of different character, such as semi-arid and mountainous regions. Migrants arrive with wind and sea currents, rivers, and migratory birds and mammals. But above all, tropical forests possess several niches that permit the permanent establishment of migrant species, and thus serve as an important refuge.

Indeed, the diversity in tropical forests is so intricate that, for all we know, there may be special biochemical and biophysical characteristics that support their unique specialization and general luxuriance. Perhaps the high temperatures and humidity accelerate genetic mutation rates within this most mysterious of all natural worlds.

Temperate forests are more simple, falling into two major categories: deciduous and evergreen. The first type occurs in warm areas that have mild growing seasons and moderate precipitation, followed by a colder period which initiates leaf fall. The most extensive deciduous forests occur in eastern North America, much of

Europe, and eastern Asia. The European forests are not generally as rich in species as the American and Asian forests. Temperate deciduous forests are much less extensive in the Southern Hemisphere, occurring only locally — in parts of eastern Australia, southern Africa, Brazil, and Chile — because of climatic barriers.

The effect of sunlight on temperate forest floors is particularly noteworthy. In the spring a variety of plants grow profusively as the forest floor obtains abundant sunlight through the leafless branches. As the trees begin to produce leaves, the light intensity on the forest floor dwindles. The plants here, however, continue to produce flowers that set seed until the leaves on the trees grow enough to close out large amounts of light.

Temperate forests are also composed of coniferous evergreens with small compact leaves protected by thick cuticles. These so-called needles prevent excessive water loss in plants such as pine, fir, spruce, and hemlock.

The kind of climate that produces these forests occurs generally in the interiors of large continental masses. Most of the precipitation falls in the summer. In winter the atmospheric air is characterized by very low moisture content, and the rate of evaporation is very low.

Forests—be they temperate or tropical—generally build up and maintain their soils for very long periods of time by means of an evolutionary feedback system. The roots physically penetrate

Land and water roads open Amazonia. Below: powerhouse nears its anchorage on the Jari, an Amazon tributary. Mr. Daniel K. Ludwig's Jari enterprise could yield practical data for ecologists and tropical resource managers who continually weigh the changing economic priorities of developing nations.

the topsoil and subsoil; falling leaves and fruits provide raw materials to the soil; and microfauna and microflora decompose these raw materials to form humus, from which the trees extract nutrients.

In temperate forests, the increment of leaf fall during the autumn and winter is so great that the decomposing agents cannot work fast enough to break down all the leaves in the spring and summer months. By the time they have just about decomposed all of last year's litter, along comes autumn and dumps some more.

In the tropics leaf fall is a daily occurrence and the decomposing agents work faster. As soon as a few leaves fall, they are quickly broken down in a way that is not always noticeable, even to a forest ranger. And just as quickly, the soil's nutrients are taken up by the roots of the trees. Indeed, unlike the temperate forests, nutrients in tropical forests are stored in the trees, not the soil. While this is now fairly well known, many people continue to assume—upon seeing the luxuriance of the tropical forest—that the soils are equivalently rich, making them prime areas for agriculture. This, sadly, is not true. Temperate zone woodlands actually have richer, more resilient soils.

Natural disturbances in the biosphere—earthquakes, volcanic eruptions, and hurricanes—affect the environment in a traumatic way. In the short term, they cause what seems to be large ecological shifts, but through the passage of time nature generally regenerates. The damage of human interference in ecosystems, though, is often permanent; massive tropical deforestation may be the most serious of our misdeeds. Deforestation in tropical zones now proceeds at a rate of 27 million acres per year, as forests are cut or converted to other uses: timbering, farming and ranching, and human habitation. Large hydroelectric dams may spring up to supply an ever-increasing population. Reservoirs for the dams flood out large forest areas.

And it doesn't stop there. Some of the machines used in the tropical timber industry of West Africa are capable of loading 500 tons per day, and during a good season just one machine can load 10,000 tons of logs

per month. Such exploitation leaves behind a tremendous amount of waste wood after the harvest of a few choice logs. Gross deforestation around the world brings large-scale erosion due to rapid runoff of the torrential rains. Hardpan stays behind after fine material runs off and contributes to the siltation of nearby waterways and lakes. Without its trees, a forest area may support large-scale agriculture for two or three years before it loses its vitality.

In addition, our relentless decimation of the world's tropical jungles is quickly removing the opportunity to inventory and understand their puzzling diversity—approximately three million different species of organisms. More than 30,000 flowering plant species occur in the Amazon Basin alone, compared with the 20,000 that comprise the entire North American flora. Tropical rain forests hold great promise for the discovery of new medicines, foods, gums, waxes, industrial oils, and insecticides, as well as ornamental indoor plants. If we continue on this careless path of deforestation, not only will the options provided by new research on the biota be closed to us, but humanity itself may be severely damaged by the loss of plant and animal life on which we so much depend yet so little appreciate.

It is useful to recall that one of the by-products of the photosynthetic process is oxygen. The tropical forests are one of the major sources of the oxygen the world breathes.

The sun, our life-giving friend, can easily change into a venomous enemy, dealing out doom to thousands of people who may be destined to perish upon brick-hard, parched soils devoid of even firewood to cook a nonexistent meal. The benefits of sunlight are being eclipsed by mankind's denudation of the tropical forests—a vast reservoir and a planetary trust. It must become the business of governments, environmentalists, and commercial interests the world over to bring such practices to an early halt.

Latent energy in vegetation explodes as jungle burns to clear land on the Jari River. A single species of pulpwood will be planted. Left, Roberto Burle Marx—landscape architect of Brasilia and other cities—works in his nursery. His conservation efforts, joined with others, can help save the endangered tropical world from the dangers inherent in deforestation. Top: old German bookplate shows our enduring Earth beset by winds of change.

Solar Rhythms

And God said, let there be light and there was light. And God saw the light, that it was good; and God divided the light from the darkness. And God called the light day and the darkness he called night. . . .
　　　　　　　　　Genesis 1:3–5

Sterling Hendricks

Day and night give the beat to which all things march. Man, animals, and plants fit into the relentless order of the seasons. Some creatures find a niche for living even at the ends of the Earth, and from Arctic to Antarctic the environment is suited for life and life is suited to the environment.

Heat is the first need for life, and it comes wrapped up in sunlight. We bask in sunlight, and though we might disagree on a cold winter's day, Earth's temperature is truly ideal for life. Venus is too hot; all the other planets are either too cold or lack an atmosphere. Varia-

tion is the law of life, evident from the poles to the equator and from the plains to the mountain tops, with creatures adapted to the many climes. The sun is beneficent and life is plastic. And at the planetary level the forms and cycles of life fit into one vast scheme.

At the down-to-earth level of individual species we recognize varied ways of coping with the annual swing of the sun, taking examples from an Aesop's fable of 2300 years ago:

On a cold frosty day an Ant was dragging out some of the corn which he had laid up to dry in summer. Half perished with hunger, a Grasshopper begged the Ant to give him a morsel to preserve his life.

"What were you doing," said the Ant, "this last summer?"

"Oh," said the Grasshopper, "I was not idle. I kept singing all the summer long."

Said the Ant, laughing and shutting up his granary, "Since you could sing all summer, you may dance all winter."

Aesop's own moral was: Winter finds out what summer lays by.

But as nature has arranged it, the Grasshopper can no more store away food in summer than the Ant can sing. The grasshopper and its kind, in reality, have as much foresight as the ant. In response to the short days of autumn grasshoppers lay dormant eggs. Trees also ensure against the winter by forming dormant buds that will grow again in the springtime. Even some mammals grow dormant. Pituitary and thyroid hormones control the seasonal changes in activity and also assure that the hibernating animal stores enough food—as body fat—for winter.

People couldn't hibernate if they wanted to:

Arctic mother and cub huddle on pack ice off Alaska, left. Solitary wanderings of the great polar bears cover hundreds of miles, a fraction of the Arctic tern's epic 22,000-mile migration. Many birds shuttle with the season, breeding in the north and wintering in the tropics. Right: rice paddies flank a Philippine mountainside, a sign of man's own homage to the sun that orders all life.

hormone action at work again. Man simply cannot endure hypothermy, or a drop in his internal temperature below 37°C. And partly for this reason the hominid line—including man—developed in the tropics of Africa. Our distant forebears spread to the temperate regions only after learning about clothing, shelter, and the seasons.

Indeed all creatures have ways of best using the 0° to 50°C range of temperatures and of avoiding or enduring the harsher extremes. We manage by being warm-blooded (or endothermic, which means heat from within) and using clothing and shelter for insulation. Food for metabolism is the source of the internal heat.

Food itself represents sunlight that has been transformed into simple chemical fuels. Various plant and animal products can be burned in open air to warm man's shelter, or even cool it, when comfort is the desire. The primitive Eskimo in a harsh environment below 0°C (32°F) cannot survive without heat from fuels: foods metabolize in his body, animal fat burns in a dish. The snow of igloo walls—a good insulator—holds heat for a long time. The Arctic hunter's energy supply ultimately comes from sun energy, as gathered by mammals that pawed the snow for dried plants or from sea-going mammals that reap a bounty from the deep.

And the sea is a shelter protecting many creatures from temperatures below 0°C. Ocean waters mix over the earth. The sea's circulation assures a temperature above 0°C with a high of about 40°C. Both marine and land mammals and many birds can develop insulating layers of fat below their skins and thus cope directly with cold.

Snakes, ants, and other animals that maintain a somewhat higher temperature than their surroundings (even though the rate of metabolism falls as the temperature drops) need heat in winter. Both snakes and bees protect themselves against low temperatures by huddling together to benefit from the heat derived from their very low metabolic activity.

Temperatures of reptiles, insects, and many other creatures we call the ectotherms (which means heat from without) remain near the value of their surroundings, their responses matched to both the day and the season. Most of them become torpid as the temperature approaches 5°C (41°F). Snakes literally can't survive at the higher latitudes, where summer temperatures in the 40s and 50s are common. Even the activities of mosquitoes and flies, the worst pests in the short summers beyond the polar circles, are limited by temperatures below 5°C. Their season can be less than a month during which they must lay eggs for survival. Packed glacial snow, while seemingly devoid of life, may actually harbor small black worms living with red algae in the melted water held between ice crystals. These associated plants and animals both benefit from their dark color which absorbs sunlight and from a highly adapted metabolism.

Heat from the sun helps maintain the rather narrow temperature range needed for life. The extreme upper limit for most animals is 50°C (122°F), about the temperature of a hot bath and as high as the sun can heat a creature in open surroundings.

Life depends on many interrelated reactions, many of which would be disrupted by temperatures higher than 50°C. And to be effective, these reactions must operate in a balanced way. A jogger's metabolism goes up and extra heat is given off. A walker stays cooler. Enzymes do the actual regulating, and each enzyme regulator has a limited temperature range for highest effectiveness. The enzymes are proteins and most become unstable in the region of 50°C, tending toward coagulation—as an egg's protein congeals when it is poached.

Only water in liquid form can sustain the chemical reactions needed for life, and water solidifies at 32°F. But even near the poles, ice thaws for short periods in the long daylight season, and at the polar circles the sun warms the land enough to maintain an average temperature greater than 5°C for about three months of the year, a level adequate for some tree growth. In fact, the Arctic Circle generally marks the limit of forestation.

Life responds to seasonal extremes in rhythmical ways, attuned to cycles of day and night. The sun's rhythm helps creatures anticipate seasonal changes and prepare for them. Shortening day length is often the "trigger," touching off a series of both internal and external reactions. We see the consequences in migration, reproduction, hibernation, dormancy, diapause, flowering, and photoperiodism. The internal rhythm is ubiquitous in all living things.

Formal scientific recognition of biological rhythm came first from an awareness of leaf motion, plants having long been known to change the orientation of their leaves from day

Man, animals, and plants share a unity of response to the cycles of day and night and the seasons. Examples from the story appear in the diagram, opposite. In many cases, sunlight triggers innate, internal rhythms that help motivate life forms. Above: 19th-century print depicts America's night-blooming cereus, or moon cactus, in full nocturnal splendor. Like many desert plants, it reserves activity for night hours.

to night. In 1729 J.J. deMairan, a French astronomer, noted that plants placed in continuous darkness maintained the daily change of leaf orientation. This diurnal movement was studied in detail during the next two centuries. Charles Darwin and his son Francis took up the work and wrote a book on *The Power of Movement in Plants.*

In 1918, H. A. Allard and W. W. Garner of the U.S. Department of Agriculture made a fortunate observation while working with tobacco plants. They noted that a specimen in one of their test plots kept growing while others matured at a smaller size. Brought to the greenhouse, the overgrown plant soon flowered, as did young ones of the same variety. Tests by Allard and Garner did not result in a more productive crop variety as they had hoped, but they went on to determine that tobacco plants require the stimulus of short days to flower.

They coined the term photoperiodism for such responses to light, and eventually found that poinsettias, chrysanthemums, soybeans, and many other plants flower best when the days are short and the nights are long. Asters, barley, wheat, and some other plants flowered only when the days were long, during spring and early summer.

Allard, also interested in animals, knew that many mammals, birds, and insects migrate and reproduce depending on the season. He reasoned, in analogy with plants, that the animals were responsive to the changing length of the day. This possibility was tested in 1925 by W. Rowan, a zoologist in Edmonton, Alberta. Rowan kept juncos and crows in outdoor cages during the early winter but simulated the approach of spring by adding five extra minutes of light each successive day. The banded crows were released in midwinter, and their bands were recovered chiefly from the northward regions. A trapper hundreds of miles to the north, on hearing a crow caw in the depth of winter, shot it, to prove his sanity.

Retained in cages, the juncos developed sexually during the increasingly long yet still frigid days. The day and night length was right for them even though the temperature was wrong.

Many of the invertebrates also reproduce and migrate depending on the day length. S. Markovitch, starting his career as an entomologist, was inspired by the findings about plants. He went on to experiment with specialized aphids that hatch from over-wintering eggs on apple trees. Several generations later, they migrate to nearby dock weeds where they live and reproduce, hatching a flightless form by the long days of summer. Autumn brings winged aphids which migrate back to apple trees and lay dormant eggs, thus starting the cycle anew. The apple aphid and its kind are perfectly attuned to the season.

In the early 1930s, a German, Erwin Bunning, repeated and extended the experiments of two centuries earlier. He realized that the movement of the leaves depended on some type of changing function within the plant, and that evolution endowed plants and animals

Keyed to the sun's cycle, gigantic blue whales navigate in plankton-rich polar seas, moving toward tropical waters as pack ice forms. Also marching in step with solar rhythms, aphids feed on a flower which itself sprouts, blooms, and dies in the familiar seasonal pattern.

146

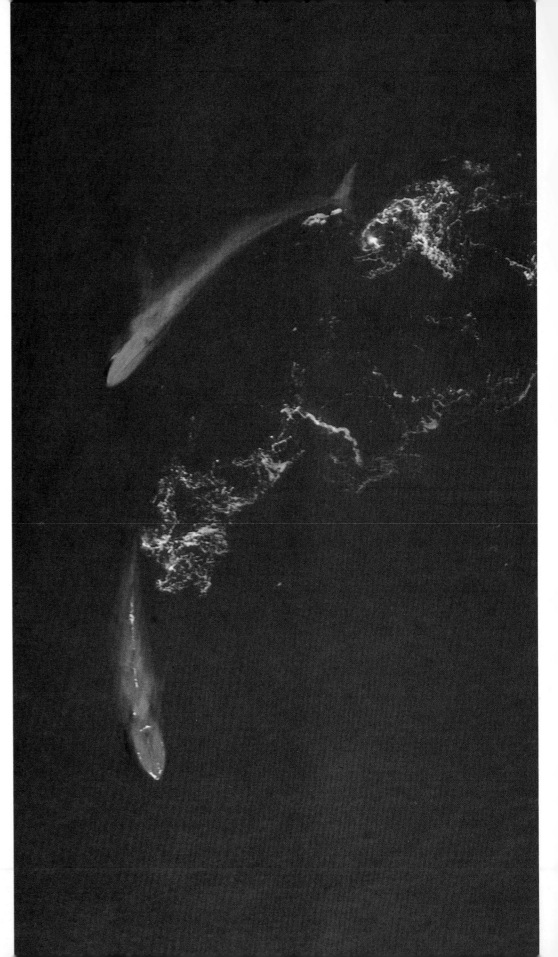

with the ability to maintain a sequence of physiological changes that reflects the alternation of days and nights. While this notion appears vague on first hearing, it has the merit of being testable. Almost any repeated display can be examined experimentally. Bunning, inspired by the outcome of numerous studies, coined the term "biological clock."

The new concept of an internal clock that moves in time with the sun inspired much work. Mice, cockroaches, and many other small nocturnal animals are well suited for tests of the endogenous rhythms, cycles of chemical and physical responses initiated from within. For instance, with adequate food and water animals can be held for weeks in darkened cages. The cages have running wheels, or squirrel cages, with a method of recording the rotation of the wheel. The records of activity show cyclic repetition, with periods between greatest activity often repeatable to a few minutes.

The endogenous basis for rhythmic displays is well expressed in the hormonal controls of animal and plant life. With chickens, for example, the change from darkness to light stimulates hormonal activity. Each hormonal signal induces the release of the egg from the ovary. These steps take time and, over all, the cycle of events in the hen is somewhat more than 24 hours; consequently eggs are laid later and later on successive days until the total increase in time is about eight hours. All the eggs laid successively form a clutch, and, having completed one, the hen skips a day before starting anew. In the wild, clutches usually contain fewer than eight eggs.

The internal rhythms of birds and mammals interact with the cyclic change of day and night to anticipate a coming season. Ptarmigans, weasels, and snowshoe rabbits change their coats from the variegated colors of summer to the white of winter. Hormonal changes cause fat formation to increase during the summer period of plentiful food, and so provide extra energy for migrating birds and insulation for nonmigrating animals.

Man is diurnal rather than seasonal in his functioning, yet his sequencing depends on glandular secretions. In a less obvious way the hormonal control of his functions goes on without his conscious effort. He wakes and he sleeps. Cyclic changes are measurable in terms of the release of hormones from his endocrine system. The hypothalamus, pituitary, adrenal,

147

and thyroid glands are involved, as well as the pancreas and gonads. Compounds related to cortisone, as an example, show diurnal variations in the blood stream. The hypothalamus normally changes rhythmically, and becomes out of phase with the day when one travels rapidly over long distances. Jet-lag results.

But both people and plants have direct external responses to the sun. To study plant processes, scientists vary the amount and quality of the light reaching plants. Investigators discovered that short-day plants, the soybeans and cockleburs used by Garner and Allard, fail to

plants for recognizing the surface of the ground and in becoming green. The lengthening of the stem in darkness allows a seed to be planted at a reasonable depth. When the shoot from the germinating seed breaks the surface of the ground the whole pattern of growth changes: there is a dramatic, total plant response to the alteration of a minute ingredient of its chemical make-up, triggered by light from an otherwise insignificant component of the spectrum of sunlight.

The pigment, which has been isolated, is a blue protein with a chromophoric group that

From arctic to the tropics, animals—and people —have assured their survival through migration. Here, wildebeests trail across the Serengeti plain in Tanzania in their search for plentiful food and water. The arduous migration is taken in stride, the group adapting en masse to avoid extinction. Right: female katydid lays her eggs in the tissue of an oak leaf at summer's end. She shares the

flower if the day length is extended with dim light or if the nights are interrupted with light for a few minutes. Red light is most effective. It turns out that a red-light absorbing pigment, namely a blue-colored one, is in control of the flowering. If the exposure to dim red light is followed by a short exposure to far-red light, just at the limit of vision, flowering occurs. This reversibility indicates that the controlling pigment exists in two forms which can be interchanged by light.

The reversible change of the blue pigment can be used to test whether other responses of plants to light or dark are controlled by it. Young plants can be grown in darkness, for example, and sprout up with long stems, small leaves, and a pale yellow color. A short exposure to red light increases the rate of growth of the leaf and decreases that of the stem. The effect is reversible. The response is universal among

absorbs red light. The reversible change occurs in the chromophore, the pigment being associated with one or more cellular membranes in the plant. The change in the chemistry of the chromophore influences the action of the membranes that in turn control the plant's metabolism.

Something rather similar happens in animals. Rhodopsin, or visual purple, is the pigment in the retina of the eye which is sensitive to red light and undergoes chemical changes when light strikes it. The pigment is composed of protein with a small associated molecular group, the chromophore again, that absorbs the light. Small changes in the protein, the "opsin" of rhodopsin, give three main regions of absorption of visible light by the chromophore. These three regions of absorption are the basis for color recognition by the eye. Light absorbed by the rhodopsins leads, through an elaborate

fate of the grasshopper, which also perishes in the frost. Each species has its own particular way of overcoming severe changes of climate. The katydid, for instance, lays eggs that stay viable through the winter, then hatch in spring.

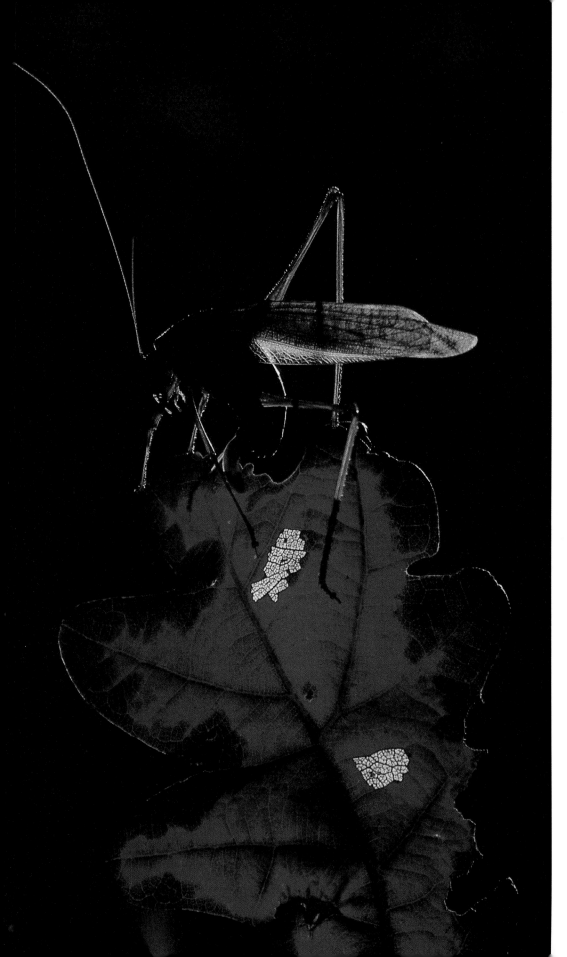

process, to an electrical signal that is conducted to the brain by the optic nerve. The brain is the control center mediating the response. The structure of eyes differs widely in the different creatures, but the basis for vision is the same.

It's no accident that the reversible chromophore reaction occurs in both plants and animals. Radiation absorbed from the visible spectrum of sunlight contains sufficient power to raise the energy level of an electron in a colored compound; and that energy can drive reactions that otherwise would not take place. Such photochemical reactions enable us to see and, through vision, to recognize color. Not only do men and animals sense the environment through the chemical reactions of eyesight, but similar pigmental compounds are also the basis for photosynthesis, the plant's most important energy-fixing process (pages 126–133).

Through such light chemistry most living things synchronize their internal cycles with those of the sun and the rotation of the Earth. For instance, birds migrating over long distances use both sight and the phase of their endogenous rhythm to aid navigation by the sun and the stars. Sheep come into heat during the season of long nights, breeding in the autumn and lambing in more favorable springtime weather. Caribou have summer and winter ranges separated by as much as a thousand miles. Their stimuli for migration, reproduction, and change of coat are the changing day and night lengths rather than the temperature. The internal hormonal controls, although not studied in the caribou, surely depend on endogenous changes in a manner similar to birds. This universal control system appears to have a survival value, as length of day changes at a constant rate, while temperature fluctuations are fickle—Indian summer, for example, lingering on just before hard winter frost or storm arrives with lethal swiftness.

From these examples and many more we can extract a simple and powerful principle at work. Man, animals, and plants have a unity of response to the cycle of day and night, despite their diversity and complexity. The unity is in the need for the heat of the sun and of the ultimate control by light of the internal biological rhythm. In the Songs of Solomon we read:

For lo! the winter is past; the rain is over and gone; the flowers appear on the earth; the time of the singing of birds is come, and the voice of the turtle is heard in our land.

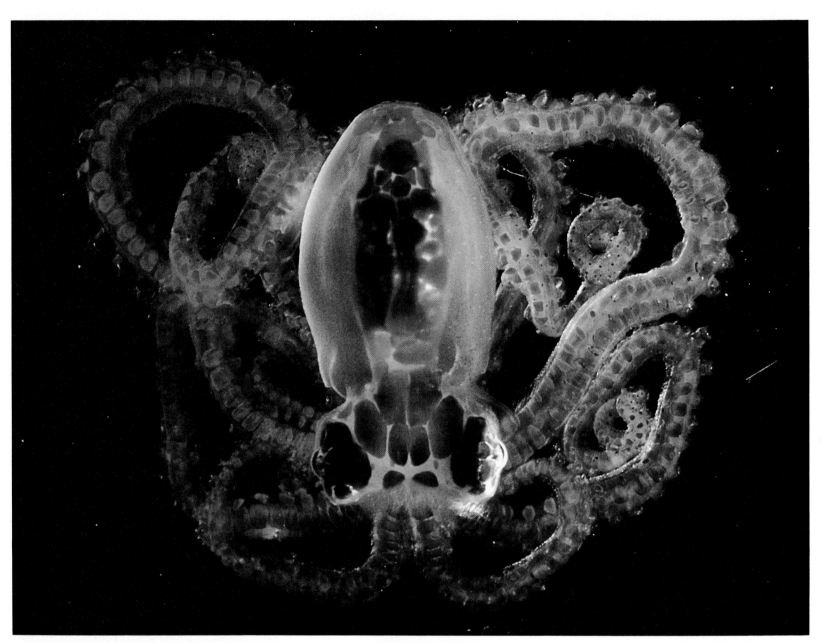

The Sun
and
the Sea

John E. McCosker

There is general agreement among my oceanographic colleagues that Earth is a misnomer for our planet. More appropriately, it should be named La Mer, in honor of that ionic liquid that bathes nearly 70 percent of its surface. Imagine, as do certain island cultures, the romantic imagery of El Sol, the breadwinner of the solar system, in an energetic marriage to La Mer, passing along his life-supporting forces to her through photosynthesis. John Isaacs of the Scripps Institution of Oceanography has roughly calculated that of each million photons which strike our planet,

only 90 are absorbed into the net production of basic food: nearly half of these become energized into land plants and the rest contribute to the growth of the unicellular plants which make up the meadows of the sea.

We embark now on a sea voyage through the verdant pastures of her surface to her dark and monotonous abyssal depths. We will stop at various levels along the way to investigate the adaptations that plants and animals have made with regard to the amount and kind of available light.

The oceans and seas have an average depth of

four kilometers (2.5 miles), yet it is only the upper 200 meters (600–700 feet) or less of clear oceanic water that is illuminated with light of sufficient intensity and quality to allow photosynthesis. The light in this layer (the euphotic zone) depends on the amount of solar radiation striking the sea's surface, the amount reflected back to the sky, and the absorptive materials within the water which reduce its transparency. Ultimately, all light energy is thus scattered, reflected, or absorbed. Absorption and conversion of sunlight by plants into high-energy compounds supports all life on Earth. Marine biologists measure the rate of this organic turnover, called primary production, and have learned enough within the last several decades to realize how much they still don't know.

Seawater differentially absorbs the component wavelengths of light. (Recall throughout this discussion that the color one observes is the particular spectrum reflected by the object; other wavelengths have been absorbed.) Infrared, red, dark blue, and ultraviolet wavelengths are absorbed within the first few meters of oceanic water, so much so that at ten meters (33 feet) light is predominantly blue-green in color.

Perhaps if marine biologists were fitted with seawater-simulating contact lenses, they might better appreciate what life is like in a world without red hues. Most shallow-water fishes have a retina complete with cones for color viewing and rods for detailed black and white vision. As one might suspect, the optimal color reception of each species happens to coincide with the spectrum of light which penetrates to the depth the creature inhabits.

In their evolution, marine plants have, like fishes, finely tuned their light-gathering pigments to the spectra available to them, not for vision or camouflage, but for photosynthesis. Among the photosynthetic pigments of phytoplankton (tiny free-floating plants) are carotenoids and chlorophyll; under the microscope, these plant cells appear green because of the chlorophyll-containing chloroplasts which

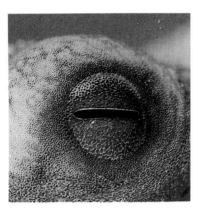

Left, a baby octopus dilates its chromatophores or pigment cells. Under water, its colors appear as shades of gray, providing camouflage as well as indicating octopus moods. Even the highly developed octopus eye, above, sports chromatophores. Top, painted dolphins from Knossos on Crete. Such marine mammals must live at the sunny surface of the sea, but may dive to eternally dark zones more than 1000 meters (3000 feet) down to feed.

capture red and blue light and in turn reflect the green. Absorption peaks of various attached and planktonic algae nicely coincide with the kind and intensity of available light.

The large attached algae of the sea, collectively known as kelp or seaweeds, are limited to the narrow coastal zone, living in 100 meters (330 feet) depth or less. As previously mentioned, the longer, red wavelengths penetrate only a few meters, so that complimentary colors are absorbed and utilized by the plants. A colorful vertical zonation exists beginning with the green sea-lettuces and their relatives which live in the shallowest water, beneath which the red and the brown algae live. Healthy kelp sprouts more than a meter (three feet) a day, making it the world's fastest grower.

Just as the attached kelp must grow in shallow water to be within the sunlit zone, so free-floating phytoplankton must stay near the sunny surface, even when they are swept over great abysses. The depth of light penetration is a key factor influencing oceanic productivity, a matter of considerable importance to us humans, who are after all the final predators, the top of the food chain. Energy ultimately derived from sunlight is passed from link to link of food chains of varying lengths and efficiencies. The efficiency of a food chain decreases as the number of links increases. Consider a land-based chain: sunlight to grass to cow to man; a football field's equivalent of grass becomes one hamburger. The oceanic food chain usually has four or five steps, such as sun to phytoplankton to zooplankton to krill or small fish to tuna-sized fish to fisherman. At each transfer 80 to 90 percent of the energy is lost to animal maintenance; thus only 0.1 percent of the original solar energy ends up in the fish dinner which we consume.

Not coincidentally, the greatest vertebrates of the sea eat lower down the food chain. The main exception is the sperm whale. Other great whales and the greatest fishes, the whale shark, basking shark, and megamouth shark, eat shrimp-like krill or smaller plankton which in turn have fed on copepods or diatoms. Thus by

their dietary preference, these oceanic giants are 10 to 100 times more efficient at harvesting the productivity of the sea than they would be as fish eaters.

Until recently, oceanographers towed plankton nets behind their vessels to discover the large-scale variations in the amounts and kinds of these drifting microscopic organisms, at each locale and in each season. Now, sensors aboard satellites provide the information. The seasonal and latitudinal variations of plant life are significant because of the fisheries which rely directly or indirectly upon the local phytoplankton.

These tiny life forms which en masse support the food chains of the sea, are a far cry from the large, foul-smelling kelps and other algae that a landlubber generally associates with a wave-swept shoreline. Except for regionally or sporadically concentrated blooms that discolor the water or create nighttime luminescence, phytoplankton are inconspicuous to the naked eye. They include the coccolithophorids, the golden-brown algae, the diatoms, and the dinoflagellates. The latter two are the most significant.

Diatoms are seasonally found in incredible abundance both in open water and in the seabed. They are an important food source for the filter-feeding animals as well as minute, bottom-dwelling shrimp-like creatures. Diatoms reside within exquisitely sculpted little silicate shells, formed of two close-fitting halves reminiscent of a pillbox. Although they are of basic significance in the oceanic food chain, it remains a mystery how these creatures move and how they regulate their buoyancy. Dinoflagellates, the second important group, are motile, having two flagella, and are either naked or possess a light, nonsiliceous armor. Their sporadic abundance in coastal waters can cause red tides—densities of organisms so great that they color the water with their pigment—which in turn can render shellfish inedible during summer months because of the dinoflagellate toxins they concentrate.

The most productive fish-producing areas of the world are, naturally enough, those where primary productivity is highest and the food chains are the shortest. These oceanic cornucopias are called upwelling regions. They occur primarily along the temperate and subtropical western continental margins (including California, Peru, northwest and southwest Africa), the Arabian coast, and Somalia. There, due to the

rotation of the planet and its resultant gyral effect upon moving water—the Coriolis effect —nutrient-rich deep water replaces nutrient-poor surface water driven offshore by coastal winds. These upwelling zones comprise but 0.1 percent of oceanic water, a combined area smaller than California, but they are responsible for more than half of the entire world's fish catch.

This excessive dependence upon such small portions of the sea for harvesting raises serious questions. What if the upwelling were to stop? It does, occasionally, during a periodically

occurring phenomenon called *El Niño* in Peru, in honor of the Christmas child. El Niño is no holiday gift, however, for when it occurs, cold, nutrient-rich water is replaced by warm nutrient-poor surface water. In 1972 El Niño—and overfishing—resulted in the collapse of the world's largest fishery. So the sun, in spite of its willingness and warmth, cannot prime the oceanographic productivity pump unless the necessary nutrients are there as well.

The bays and coastal zones, 10 percent of the ocean, are twice as productive as the open sea. In rough figures, each square meter of oceanic water produces approximately 50 grams of carbon each year, as compared with 100 grams from coastal zones. An active upwelling region takes the honors, and can produce nearly 300 grams. Nearshore zones are particularly productive of key nutrients from the continental margin. The giant kelp beds of temperate California and Peru and the coral reefs of the Caribbean and tropical Pacific are as productive as Montana wheat fields, but why? Because the nearshore tides and currents constantly move the nutrients across the rock and coral reefs— the forests of the sea.

The structures of coral reefs are composed of

Sunlit zone
0 to 200 meters

Twilight zone
to 1000 meters

Benthy-pelagic zone
below 1000 meters

the skeletons of myriads of once-living coelen-terates, whose polyps secreted the calcium carbonate substance upon which the living polyps build in turn.

In recent years, invertebrate zoologists have discovered that the coral polyps (animals) can-not perform this architectural triumph alone but depend heavily upon symbiotic algae that live within their tissues. Sunlight is captured by the blue-green zooxanthellae live-ins, which release oxygen through the photosynthetic reaction, and the oxygen is used, along with an occasional meal of algae, by the growing coral. For that reason, coral growth is limited to shal-low transparent seas.

Usually the first reaction of a surfacing nov-ice snorkler is "Why are coral reefs and their fishes so colorful?" More precisely, how can they afford such gaudy bodies if, as we are told, a predator lurks behind each branching coral? No simple answer exists, but one may safely assume that the colors serve both as camou-flage and as signals. The bizarre shapes and colors of both the coral and the fish provide such a confusing background so that a predator has difficulty zeroing in on its prey. Further-more, the conspicuous and distracting color patterns of the fish themselves disrupt the tell-tale outline that is important to the search image of a predator. But what is the communi-cative value of camouflage coloration if the viewer can't see color, or even differentiate hues? Apparently many fish can see color.

The deductive evidence for color vision in fishes comes from many sources, including studies of the microscopic anatomy of the ret-ina, laboratory experiments with living fishes, and observations by the many freshwater fish-ermen who have learned that on certain days the Quill Gordon artificial fly is preferable to the Ginger Quill on a sunlit trout stream.

Most shallow-water fish have retinas con-taining both rods and a variety of cones. Deep-sea fish, with few exceptions, have a pure rod retina; a variety of colors, you will recall, is unavailable at depth. Mid-depth fishes are sen-sitive to shorter wavelengths, the blue wave-lengths that best penetrate clear ocean water. This shortwave shift has been made by all the vertebrates that live and swim in that limited color band, including the whales, seals, bony fishes, and sharks.

A few optical exceptions exist in the deep ocean, for instance a black dragonfish which both emits and receives red light. Since most of its midwater compatriots are color blind, pos-sessing only rods, such a red-emitting predatory species in effect possesses a "sniperscope" that its prey are oblivious of. That is, the black drag-onfish can both illuminate its prey with emit-ted red light and see it in that wavelength of light.

Without the protection of a reef or the dark-ness of depth, what is a helpless sardine to do for protection in open water? The survivors are those that natural selection has favored with schooling behavior and adaptive two-tone coloration. Sunlight penetration in clear near-shore and oceanic waters has necessitated a camouflaged livery to protect such potential prey and also to disguise an approaching preda-tor. Nearly all free-swimming open-water animals are "countershaded"—they are gener-ally darker on their dorsal surface, lighter along their ventral surface, and silvery in side view. Such a coloration renders an animal nearly invisible from any angle. A silver-sided, blue-backed sardine is nearly invisible when seen from above by a diving pelican, which sees only blue below, as well as to the equally counter-shaded tuna beneath it which looks up hungrily to the surface and sees only the bouncing light penetrating through the sea surface.

Animals in deeper water, at 500 meters (1600 feet) for example, are illuminated enough by downwelling light that they appear silhou-etted against the distant sunlit surface. Early anatomists recognized the function of the large up-turned and telescoping eyes of midwater fishes. Consider the bizarre barreleye, a mid-water, hand-sized fish which lives in depths of 200–600 meters (650–2000 feet), constantly looking upward to detect the silhouette of prey, predator, or another barreleye of the opposite sex. As 20th-century oceanographers descended to the nearly lightless depths in titanium spheres and plastic bubbles, they observed attack-ready lancet fishes in a curious oblique posture, looking upward for the passing silhou-ette of their next repast.

Hunting is complicated by the evolutionary gambit of potential prey species which counter-shade (or, more accurately, counterlight) them-selves with a series of dimly glowing lumines-cent organs arranged along their ventral surface from chin to tail fin, producing a weak blue-green glow that is equal in intensity to the downwelling sunlight.

The deeper midwater zones, called the mesopelagic—arbitrarily defined as 200—1000

153

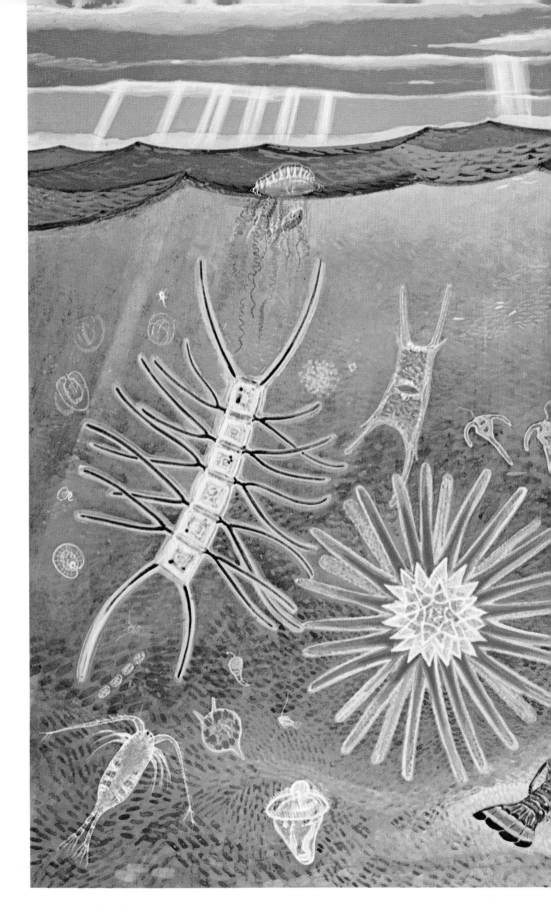

The areas where marine life abounds are rich in a combination of sunlight, nutrients, and highly oxygenated water. Such areas occur most frequently on continental shelves and atolls, and where upwelling brings cold, nutrient-charged water to the surface. At the bottom of the food chain are the phytoplankton, single-celled freefloating plants that convert sunlight and inorganic nutrients into their own structures. These plants and the equally tiny animals that graze on them in turn provide food for everything else. Although the oceans' depth averages about four kilometers (2.5 miles), only the upper 200 meters (600–700 feet) are sufficiently sunlit to support photosynthesis. The life of this "euphotic" zone is depicted in the accompanying illustration. Prominent in the left foreground are various kinds of plankton: tiny armored plants called diatoms; dinoflagellates, free-swimming plant-like organisms; minute crustaceans called copepods; larva of marine worms; and tiny jellies. Above is a Portuguese man-of-war; beyond are bluefin tuna preying on herring while flying fish take off. A spiny lobster creeps across a tropical seabottom near coral and undersea grasses, home for sea urchins, triggerfish, grunts, and the barracuda. In the distance, a dusky shark patrols near a green turtle below a school of mackerel. In colder seas, a brown pelican dives on frantic anchovies, as a mighty fin whale harvests vast quantities of shrimp-like krill.

154

meters (650—3500 feet)—and bathypelagic—below 1000 meters—contain some of the most bizarre and curious of all living creatures. In the near absence of light, and therefore of photosynthesis, meals are limited to what rains down from above or to the fellow sitting at the dinner table with you. Such conditions have resulted in toothy maws like that of the fangtooth, adapted to enjoy a large and struggling, but infrequent meal.

Freshly captured lantern fishes and mesopelagic squids and shrimp have been placed in chambers which precisely simulate the light of their accustomed zone. As the experimenter increases or decreases the chamber's overhead illumination, the fish, shrimp, or squid increases or decreases its ventral lighting to countershade itself. This is laboratory verification of what probably occurs when a moonlit night becomes darkened by cloud cover or the individual changes depth.

Using another feeding strategy, some mesopelagic creatures make nightly feeding forays into the upper layers and return to the dark safety of the mesopelagic at daylight. This vertical migration is based upon the zooplankton (primarily crustaceans) that ascend from the depths at sunset to feed upon phytoplankton, and the lantern fish and squid which follow them up to the surface to feed upon them. Their dark milieu is modulated by moonlight; they all rise higher on moonless nights. This phenomenon is called either DVM (Diurnal Vertical Migration) or the DSL (Deep Scattering Layer), the latter term referring to the sound reflections which perplexed the U.S. Navy when sonar submarine searches first turned up mysterious daily-varying echoes. The reflective culprit, which appeared as a smudge on sonar tracings, was finally identified by marine biologists as dense schools of fishes with gas bladders that act as excellent acoustic reflectors.

In the meso- and bathypelagic layers, both life and light dramatically disappear with increasing depth. The upper mesopelagic biomass is only about one-fiftieth that of the surface. In the lightless abyssal zone, the biomass is just one-fiftieth again of that in the zone above. The abyssal plains lie below the continental slope and span nearly half of the Earth's surface. This is thus the largest single habitat on Earth, and a relatively stable one.

Dwellers of the abyss have adapted to life entirely without light. As a species that depends on vision as the primary sense, we find

Below, living examples of sea life's dependence on the sun, the Maldives' 19 coral atolls perch atop a submarine ridge that rises abruptly from the abyssal plain of the Indian Ocean. The coral builders, tiny animals, live in a symbiotic relationship with blue-green algae. To survive, they must ceaselessly build the atolls to keep the living layer of the reefs within the shallow sunlit zone of the sea. Coral structures provide a tropical habitat for such creatures as the clown fish and sea anemone at right. Other species of sea anemones, left, flourish with kelp in shallow, colder northern Pacific waters.

it hard to appreciate the increased sensitivity and use made of the other environmental receptors available to a blind creature. Abyssal animals have little knowledge of or care for the season of the year or the time of the day. The uniformity and constancy of temperature, darkness, and minimal food availability have given creatures an environment they can count on and adapt to. There is no advantage to being reflective, or red, rather than black and unpigmented. Fishes here are typically black, weakbodied, large-toothed, and small-eyed, probably more dependent upon their senses of taste, smell, and feel than anything else.

Several of the abyssal creatures have not forsaken light as a useful device, so in its absence they make their own. The ability to make light in a lightless world confers upon the luminary a unique opportunity to see, to communicate, to confuse and confound its coinhabitants. Bioluminescent organs include small photophores which act as personal billboards to announce one's species and sex, and also as lighted lures that attract incautious prey in search of an even smaller glowing meal. The lures are usually positioned at the end of an elongated dorsal fin spine which extends and droops in front of the predator's toothy maw, or at the end of a long chin whisker or tail that is dangled in front of the mouth like a tempting worm on the end of a fishing line.

Prior to 1977, it would be sufficient to say that life and the dwindling effects of sunlight end in the abyssal ooze. But a most remarkable discovery, made by oceanographers in submersibles diving along submarine ridges southeast of the Galápagos Islands, turned up an entirely different biological approach to life. Bacteria living along the edge of volcanic vents of the Galápagos Rift Zone, more than 2500 meters (8000 feet) below the surface, have evolved to become "chemoautotrophic," nourishing themselves not on the energy from sunlight trapped and converted by plants, but on the reaction with hydrogen sulfide vented from the Earth's interior. These bacteria are the basis of a food chain that includes fishes, crabs, clams, and strange coelenterates called "dandelions" by their submarine-borne observers.

For me, the excitement of the Galápagos discovery is not so much *what* was found but that *we are still able* to make such a finding. This should be a signal to the next generation of oceanographers that much remains to be seen and learned.

A famous Jesuit astronomer of the 1870s, Fr. Angelo Secchi, wrote, "Several peoples of antiquity worshipped the Sun, an error perhaps less degrading than many another since this star is the most perfect image of the Divine, the instrument whereby the Creator communicates almost all his blessings in the physical sphere." Following the distinguished Fr. Secchi's lead, we seek clues to the cultural and spiritual life of those in the past who have developed especially close ties to the sun—as our progeny may in a sun-oriented future.

Possibly the last active European sun-cult center—a Mithraic shrine in Roman London—went into eclipse some 1600 years ago. And a millennium had already passed since Bronze Age Danes paid homage to the sun through rituals involving the solar

Children of the Sun

chariot with a golden disk appearing at left. Careful examination of its surface reveals curious spiral designs that seem associated from quite ancient times with solar observances not only by prehistoric Europeans, but by Indians in North America. Independent development? A link with some forgotten religion that came across the Bering Strait with paleo-Indians? A Jungian archetype? Who knows for sure?

We turn to the experts. E.C. Krupp takes the reader on the first leg of a long journey of understanding, with a first stop in dynastic Egypt. Kendrick Frazier leads us to the New World, to explore the great plains and deserts and to share the sun culture of America's first settlers, the Indians. Here we witness a solar alignment at a Rocky Mountain "stonehenge"—the Bighorn Medicine Wheel.

With Silvio Bedini, Keeper of Rare Books at the Smithsonian, we witness the development of two techniques for telling time— first the sundial and then the gear. While sundials depend directly upon sunlight for telling time, mechanisms do the job independently. They eventually surpassed sundials in accuracy and led to advances in both solar and stellar astronomy. Valerie Fletcher of the Smithsonian's Hirshhorn Museum examines the sun and the meaning of its light as captured on the canvases of famous artists from Turner through the Post Impressionists. Ed Kiester updates history by peering beneath the human hide and the human psyche to discover the effects of light and of its absence on people today.

159

Some well chosen star, a mimic of Sirius, the night's brightest star, appears for a moment in the cool eastern twilight. It marks the night's last hour, and the time-keeping priest assigned to watch for it prepares for sunrise.

As the sky softly grows brighter than the stars, before the sun comes up, a sweet breeze brings the promise of another day. And in ancient Egypt each of those days is very much like the last.

In Egypt the Nile rises and falls. It is the pattern of life. The sun rises, and the sun sets. It

E. C. Krupp

The Sun Gods

is the pattern of time. And time, in Egypt, is not a string of events, but a rhythm, metered by the river, the stars, and the sun.

And the rhythm is the measure of life—and afterlife. Belief in resurrection and eternity are reinforced by the unending travels of the sun. Solar gods and their priesthoods, their temples and their liturgies proliferate. And today, after a million dawns have come and gone, we are still amazed and delighted by the cultural genius of the Egyptians. In the few ways we are able, let us visit their ancient valley of sun and symbol.

At Abu Simbel, far south of Egypt's ancient

160

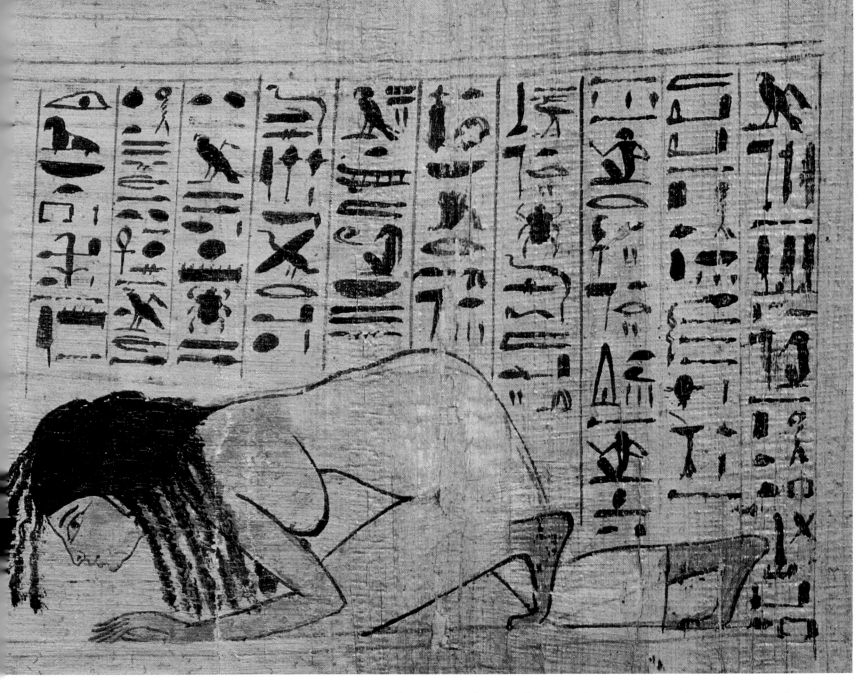

capitals, a row of baboons is carved as a cornice upon the facade of the temple of the pharaoh Ramses II. This unique rock-hewn structure permits morning sunlight to enter its innermost sanctuary only on certain, specially chosen days of the year, but the sun's *first* light is always caught by the pious baboons. They sit, paws lifted, to greet the sun.

It may seem incongruous that baboons invade the solemn precincts of Egyptian temples, but the baboon was an emblem of the sunrise. Baboons chatter at dawn, and their morning screams and antics were as much a part of Egyptian culture as the rooster's crow is of ours.

In Egypt the sun was an important component of religion. The Egyptians used the sun as a metaphor. Its cycles were the pattern for this life and the transition to the next. The Egyptians worshipped the sun as the source, or rather the inspiration, of life. It energized the world by giving it order, or rightness. Indeed, the sun made the world possible. Anything so great necessarily has many aspects, and so we find solar symbols, solar myths, and solar gods in many expressions of Egyptian culture.

A funerary papyrus from Egypt's 21st Dynasty (1085–945 B.C.) depicts a woman's soul bowing before the eye of Osiris, god of the dead. With her is a baboon, symbol of Thoth, god of wisdom and the moon, who supplies the dead with guidance for a safe passage through the underworld. In Egyptian religion, the sun made a similar journey each night, braving the perils of the dark to be reborn at sunrise.

161

King of Egypt during the Fourth Dynasty (2613–2494 B.C.), Khafre seems to stare across time at his pyramid, right, one of the three great pyramids at Giza. As the god Horus, Khafre's falcon was a longtime symbol of sky and sun. Horus later became associated with the horizon sun, Re-Harakhty. Another form of the divine sun was the Great Sphinx, built during the reign of Khafre and thought to bear his features. Below, tomb reliefs from the ancient necropolis of Sakkara show life in the third century B.C. Egypt's timeless cycles persist. The sun rises and sets, the Nile flows on, the land blooms and yields its harvest.

Khepri, the solar scarab beetle, appears on jewelry as well as on the walls of tombs and in the papyrus pages of the *Book of the Dead*. Although the baboon becomes an appropriate symbol—once we appreciate its morning routine—for the dawn, it is more difficult to sense why the Egyptians chose the scarab to represent the rising sun. To understand this we have to remember the richness of Egyptian allusion and the Egyptian delight in converging threads of association.

The scarab beetle constructs a sphere of debris and dung, then deposits its eggs within the ball and buries it. When the eggs hatch, the larvae feed on their nest egg, develop into beetles, and emerge from the soil.

To the Egyptians the scarab must have been an intriguing image. Here was an insect that created a sphere, like the round sun created at dawn. The beetle's ball rolled across the ground, just as the sun rolled across the sky. Both sun and ball entered the earth, the sun to emerge newly created the next morning and new scarab beetles to emerge, in what seemed to be spontaneous creation, from the earth. Beetles, like the sun, were imagined to be self-created.

Skeptics of this explanation of the scarab's association with the sun argue that it was the similarity between the sounds of words for "beetle" and for "coming into being" that made the beetle mean sunrise. Certainly the Egyptians were aware of this pun and delighted in such word play.

When we think of the Egyptian sun god, it is Re that comes first to mind. He is the sun personified and the world's ruler. For this reason the king, or pharaoh, was identified with Re and regarded as his son. Usually portrayed with a human body and a hawk's head, Re is the divine sun, the sun at noon. Originally his cult was centered at Heliopolis, today a suburb of modern Cairo, but his influence spread throughout Egypt as the role of the pharaoh evolved.

The course of the sun was a daily journey made by Re in his celestial boat. The path was well defined and part of the world order. It carried the sun from the east to the west, and at dusk Re exchanged the boat of the day for another that would transport him through Tuat, the underworld.

Tuat was not imagined to be literally under the earth, but was instead that realm through which the sun sailed when night was upon Egypt. Re's voy-

162

age through the 12 domains of Tuat was perilous. The greatest danger was during the last domain, in which dwelt the giant serpent of darkness, Apep. Re's combat with Apep took place just before dawn, and if Re emerged victorious, he again exchanged his boat for the boat of the day and sailed once again through the 12 domains of the day. Each of these represented an hour, of course, and together night and day totaled 24. The concept of the 24-hour day is Egyptian, and we continue its use today.

In another version, Re assumed the form of Atum at sunset. Atum was the original creator in the cosmogony devised by the priests of Heliopolis, and in the Old Kingdom of Egypt (2800 to 2250 B.C.) the Heliopolitan ideas were the nearest thing there was to Egyptian orthodoxy. Heliopolis was a major political and religious center even after the unification of Upper and Lower Egypt and the establishment of a First Dynasty capital at Memphis, about 13 miles south of Cairo.

Heliopolis is a Greek name—"Sun City"—and it derives from the Egyptians' religious name for the site, "the Dwelling Place of Re." Heliopolis was strongly associated with the sun, but its common Egyptian name, Iunu, or On, tightens the connection between Atum as the creator and Atum as the sun.

Iunu means "the Pillar," a reference to the obelisk. Like the pyramids, an obelisk is one of those antiquities that mean Egypt. The Egyptians erected these tall, tapering, four-sided needles of stone as emblems of the sun, capping them with pyramidions, or small pyramids, usually covered with gold to reflect the first gleam of the morning sun.

The crown of the pyramid of Amunemhat III (1842–1797 B.C.) at Dahshur is itself a small granite pyramid. Its east face is exquisitely carved with a pair of eyes protected by an umbrella—a winged disk of the sun. The accompanying inscription tells us, "The face of Amunemhet III is opened, that he may behold the Lord of the Horizon when he sails across the sky."

The Great Sphinx, called Re-Harakhty, or "Horus of the Horizon," was yet another form of the divine sun. As a lion with a human head, he stares eastward through the night and guards the tombs of the cemetery at Giza. When the sun winks above the horizon, it first lights the tops of the pyramids and later warms the unblinking face of the vigilant

163

*One of the ancient Egyptians'
symbols of the rising sun was
Khepri, the sacred scarab beetle,
shown at left in a detail from
the tomb of Thutmose III (1504?–
1450 B.C.). Believers pictured
divine forces pushing the
sun above the horizon and rolling it across the sky, much as
the scarab emerged from the
ground to push its dung ball
across the ground. Sunrise over
the Nile, below, reddens the
palm-shrouded site of once great
Tel el-Amarna, a city built by
the heretic pharaoh Akhenaten,
right, (ca. 1354 B.C.) and conse-
crated to the worship of one god,
the sun. Akhenaten's religion
scarcely outlived him.*

Sphinx. Re-Harakhty also was portrayed with a human body and the
head of a falcon. He was the daytime
sun, particularly at rising and setting,
and in this guise Re came forward from
the eastern "Mountain of Sunrise."

The horizon sun, Re-Harakhty, is a
hybrid, the merging of the very ancient
sky falcon with the supreme sun. The
original Horus falcon was a sublime
vision. He spread his wings, and they
were the sky. His breath was the wind.
His left eye was the moon, his right
the sun. When he opened the right eye,
the world became light. The colors of
sunrise and sunset reveal the richness
of his plumage. The speckles on his
breast are the evening's clouds.

The name Horus probably meant
"the distant one." His name and iden-

tity as a falcon made him an apt image
of the sky. Time and circumstance,
politics and religion, made a sun god of
Horus, but the eyes of Horus continued to gaze down from temple ceilings
through even the last dynasties.

Growing influence of a cult center
would prompt its priests to graft their
own patron god with the identity and
trappings of a more universally appreciated deity. This is exactly what happened at Thebes. Near modern Luxor,
this ancient city is 430 miles south of
the Old Kingdom capital of Memphis.
During the Old Kingdom, Thebes had
little influence, and the same was true
of its patron divinity, Amun. When
Thebes rose to power in the First Intermediate Period and helped unite Egypt
into the Middle Kingdom, Amun's

status increased. Like every principal god, he was credited as the creator.

One of Amun's emblems, the ram, was probably a symbol of fertility. Eventually he was solarized and became Amun-Re, one of the most successful divine combinations Egypt would ever contrive. He (and his priesthood) inspired one pharaoh after another to build his shrine at Karnak into the largest temple complex in Egypt, and some of these temples were aligned with the sun.

New Kingdom temples are not the only astronomically aligned structures in Egypt, but they illustrate well how people often have chosen to orient monumental architecture in terms of the sky. It is one of the ways people have dealt with a fundamental need for order, or at least a sense of order.

The Egyptians looked for the connections that held the world together. Through the scarab's rolling ball, the morning's gleam atop the obelisk, or the falcon's raised eyelid, the sun was credited with creation. Each sunrise was itself creation.

As creator, the sun's role may seem to be "the giver of life." The Egyptians did see the sun in these terms, but in an abstract way. The real "giver of life" was the Nile. The river's bounty and fertility was recognized in the vegetation cycle and was formalized in the myth of Osiris, the dead and resurrected god. The sun was the source of creation, not procreation. The sun did not populate the world; it ordered the world.

More obvious expressions of the order we impose on the world are the calendar and our schemes for keeping time. The Egyptians, of course, had both, and our own evolved from theirs. Most ancient peoples hammered out a lunar calendar, and the Egyptians were no exception. But early in the Third Millennium B.C. the Egyptians were also using a 365-day solar civil calendar equal to the demands of bureaucracy and business. By Roman times the solar year's edge in ordering the affairs of empire was recognized and respected. In 46 B.C. Julius Caesar replaced the old lunar calendar of Rome with a solar calendar inspired by

the Egyptians; it eventually became our own.

Mythologically as well as practically, the sun was the world's regulator. It was all the sun's game: the board, the tokens, but most of all the rules. This makes sense. The cycles of time are given meaning by the sun's regular movements. Sunrise to sunrise, solstice to solstice, the sun's behavior *is* the world's order. That is what the Egyptians meant by creation. What is created is the natural order, and the natural order is in the comings and goings of the sun.

The sun's agent in the terrestrial order was the king or pharaoh. The sun animated the universe, and the pharaoh was the conduit. He was not the king by divine right; he was divine. In

a pun the Egyptians would have enjoyed, the pharaoh was the sun's son.

It was expected the king would protect the world order, but one pharaoh upset the established structure of Egypt and refocused solar religion on the sun's actual, visible disk, the Aten. In his own mind Akhenaten was reviving the true natural order by defusing the priests of Amun-Re. Akhenaten transferred religious and political power from the old city of Thebes to Tel el-Amarna, 300 miles downriver. There he founded a new city, Akhetaten. The name means "Horizon of Aten," and the site was pristine, unconsecrated to Amun or any other god. Amarna's greatest temple was called "Viewing Place of the Aten."

New styles and symbols punctuated Akhenaten's solar revolution. An evocative form of the sun's disk began to appear in temples and in tombs. From the circular image of the sun more than a dozen rays extended downward, and each ended in a gentle hand.

In an intuitive recognition of the sun's role as provider of life, Akhenaten fashioned a concept of the sun as a universal god, one whose gifts extended beyond Egypt to all peoples and creatures. This sun was still the source of creation and the inspiration of natural order. But life was the gift for which the Aten was loved.

The true natural order must be in the eye of the beholder. Amun-Re was a sun god with political clout, and to his priests Atenism was heresy. So the worship of Aten ended with the reign of Akhenaten.

In Egypt the sun's daily course and seasonal cycle never really changed. By creating time and place, the sun gave the world structure. It made the world what it should be, and it was always what it should be. In the long run, eternity and a moment were the same.

You can sense that eternal moment in Luxor, after Nut, the slim goddess of the sky, has consumed the sun once more. Her body is the vault above us. Her face, for now, is in the west, and the sun, like a piece of candy, has slipped into her mouth.

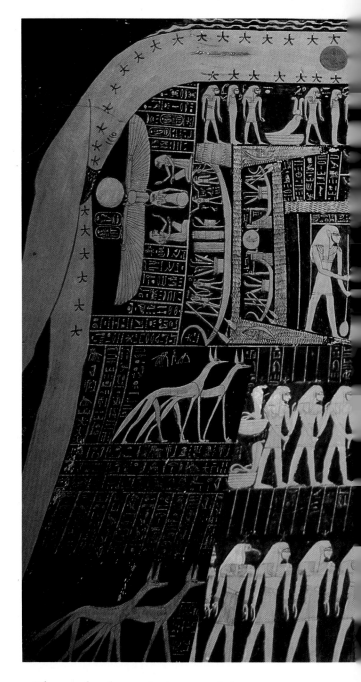

The sun has begun its journey. It is in the first perilous realm, beyond the Theban hills. Night will pass, however, and the Lord of the Horizon will reappear, reborn from the loins of the Sky. We shall have to recite prayers from the *Book of the Night*. One last hand of the Aten reaches to us over the Nile, and we pause for a moment in Luxor's evening twilight. It is, after all, the sun's specular reflection of time.

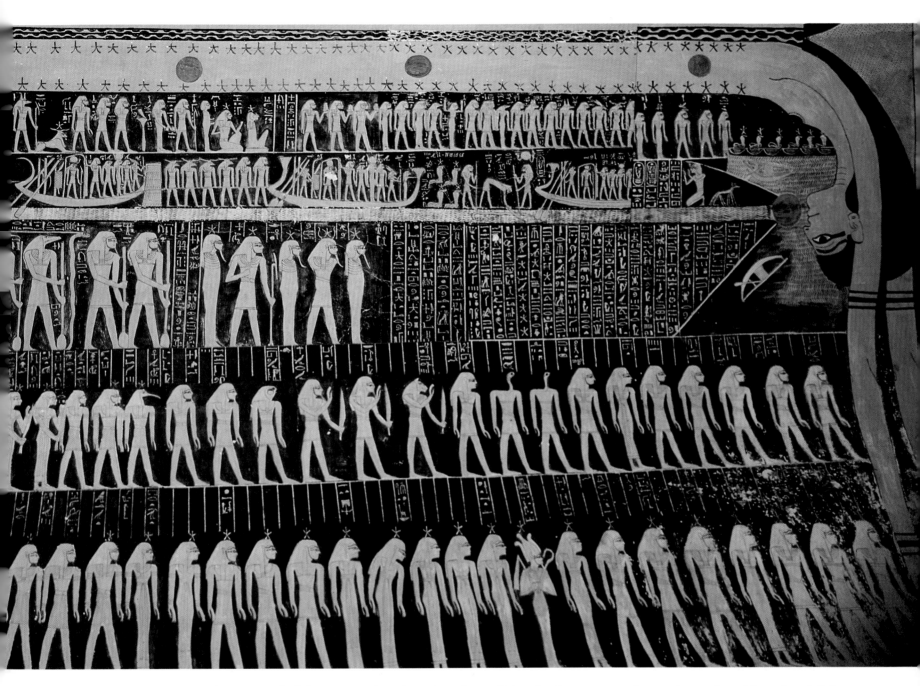

A detail, opposite, from a panel of the royal throne of Tutankhamen (reigned 1361–1352 B.C.) shows the young king being annointed by his wife Ankhesnamen. Hand-tipped rays from the golden sun-disk above them extend the ankh, sign of life and symbol of royal authority. Tutankhamen was the son-in-law and successor once removed of Akhenaten, founder of Atenism, or worship of the sun itself. During Tutankhamen's short reign, Amenism, the traditional polytheistic religion, was restored. Above: a ceiling painting from the Valley of the Kings tomb of Ramses VI (reigned 1156–1148 B.C.) depicts a section of the Book of the Hours of Day and Night. Nut, goddess of the sky, arches her body across the vault of the heavens. One Egyptian belief held that the sun was swallowed by Nut at sundown and traveled through her body during the dark hours, to be reborn each morning as the golden winged disk of the sun. The small figures represent either gods or personifications of the hours. In this reproduction, the painting of Nut has been shortened so that both ends of her long body can be seen in detail.

The dawn sun during the time of the summer solstice appears through an astronomically aligned port at New Mexico's Casa Rinconada. Rays enter a ceremonial center of the long-deserted Indian community and touch a niche in the opposite wall. *Right:* sun symbol of the Acomas.

Now this day,

My sun father,

Now that you have come out standing to your

sacred place,

That from which we draw the water of life,

Prayer meal,

Here I give to you.

Zuni offering to the rising sun

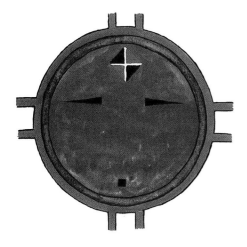

Western Horizons

Kendrick Frazier

Most of us exist smugly in our homes and offices, thoroughly insulated physically and psychologically from any real relationship with the sun and its cycles. We have our clocks for timekeeping and our calendars for marking the date. What need have we to mark the sun's daily or seasonal progressions or to integrate our lives with its cycles?

Not so the Indian peoples of America, past and present. The sun, like the land, is central to their lives, an object of reverence. Its warmth brings life, and its cycles lend a structure and an order to societies so rich in tradition.

Among the Pueblo Indians of the southwestern United States, the sun remains one of the primary deities and regulators of ceremonial life. Among the Zuni the very word for life is *tekohanane*, "daylight." Among the Hopi, the sun is highly revered, and the aid of the sun is sought in keeping the Hopi course straight. Its role as the one who "keeps the ways" is as important as its generative power. It can also have a caretaker role; thus this invocation:

We will look to our father the Sun who travels above us every day taking care of all of us, and it is he who is the highest, and in all of our religious ceremonies we take care of him in our own way, so that he will continue to perform his duty in taking care of our life on this land.

Through its seasonal change of positions, the sun also provides the coordinate system for defining the Hopi concept of Sacred Space, which, together with Time, orients man in an absolute frame of reference in relation to the universe. The sacred coordinates of the Hopi are defined by the four extreme positions of the sun on the horizon: the sunsets and sunrises at winter and summer solstices. These four horizontal directions, based on the seasonal progression of the divine sun, are joined by the two vertical directions, Above and Below.

The origin of this sun-based coordinate system is symbolized by the *sipapu*, a covered hole in the floor of each underground *kiva*, or ceremonial room. It represents the primeval sipapu, through which members of the clans first emerged from the underworld. Thus a vertical axis passes through the kiva from the Below to the world of the Above. These zenith and nadir directions are also related to the sun: its daylight position above once each day, and its nighttime position shining up from below to brighten the underworld. The six directions provide an absolute orientation that fixes Hopi Sacred Space—although since each kiva has its own sipapu and therefore its own center of Sacred Space, the Hopi concept of space has a multiplicity that ordinary space lacks.

The six directions permeate every phase of Hopi thinking and life, and lend meaning to the strongly held concept that one's village is the center of the universe. Associated with each of the directions are all manner of other things, such as colors, birds, flowers, animals, and plants. The color yellow, the mountain lion, and the oriole, for example, are associated with northwest; blue, the bear, and the bluebird with southwest. So the six directions not only provide the Hopi and the other Pueblos with a system of orientation in space but also define relationships with lesser deities and many aspects of the natural world. One can begin to see here the way Indian thought is inseparably linked with the land and the sky.

All of us are aware to some degree that as a result of the inclination of Earth's axis to the plane of its orbit, the sun's path across the sky changes throughout the year as the Earth revolves about the sun. That is what gives us our seasons. We are aware especially that the days are longer in the summer and shorter in the winter. And we may have noticed that the reason for this is that the sun is much higher in the sky in summer and very low in the southern sky (from the northern latitudes) in winter.

I've always known that happens, but it never had an impact on me until I moved to New Mexico several years ago. For the first time in my life I lived where I had a clear view, unimpeded by trees or man-made structures, of the entire western horizon. The seasonal progression of the sun was remarkable, with the setting sun traveling over a vast horizontal distance between the summer and the winter solstice.

Last summer, having become interested in these things, I decided to chart this progression in the week or so before and after the winter solstice. I carefully noted the position of the sun against the distant mountains each day at sunset and was astonished to find that the

change was noticeable from one day to the next.

Of course this was a trivial little exercise for my own benefit. Indian peoples throughout America have always known these facts, and their solar observations form the basis of their calendars, which in turn structure their agricultural and religious and ceremonial lives.

In 1879, Frank Hamilton Cushing, a 22-year-old Smithsonian Institution ethnologist, came west from Washington to spend three months observing the Zuni Indians of western New Mexico. He ended up living with them for four and a half years, learning their language and culture and joining their Priesthood of the Bow.

In a famous passage in his account of those years at Zuni, Cushing described the sacred role of the Sun Watcher:

Each morning, too, just at dawn, the Sun Priest, fol-

Wax on a board was warmed and softened in the sun, then colorful wool yarn pressed into it to create the sun plaque above. It is the product of Mexico's Huichole Indians, a group that affirms the power of the sun through its handiwork and ceremony. Right: feathers held high, a member of the San Ildefonso Pueblo in New Mexico makes a spiritual offering at dawn, as shown in this historical photograph.

lowed by the Master Priest of the Bow, went along the eastern trail to the ruined city of Ma-tsa-ki, by the riverside, where, awaited at a distance by his companion, he slowly approached a square open tower and seated himself just inside upon a rude, ancient stone chair, and before a pillar sculptured with the face of the sun, the sacred hand, the morning star, and the new moon. There he awaited with prayer and sacred song the rising of the sun. . . . Then the priest blesses, thanks, and exhorts his father, while the warrior guardian responds as he cuts the last notch in his pine-wood calendar, and both hasten back to call from the house-tops the glad tidings of the return of spring.

While the Sun Priest has the official duties, others make solar observations from their own houses, our equivalent of checking the calendar on the wall:

Hopi sun watcher helps his people organize their work and worship by observing a cliff calendar, such as the one below. The time reckoning arrangement works because the rising sun at different seasons cuts the cliff line at different spots. In spring and early summer the sun rises farther to the left each morning. The direction is reversed after the summer solstice. First sun at right moves

timing is determined largely by solar and lunar observations. Star observations govern the exact timing of many events within these nighttime ceremonies. The celestial body that regulates a given ceremony is observed daily as the ceremony approaches, usually by the chief of the specific religious society in charge. When the sun reaches a particular point on the horizon or the moon enters a particular phase, a "smoke talk" is called and the chief and the leading members of the society decide the date of the ceremony. The ceremony then begins four or eight days after the announcement, and lasts four or eight nights. (There is a numerological emphasis on the number four and its multiples among the Hopi, based upon the sun-delineated four horizontal directions.)

At least three of the primary Hopi ceremonies, as well as the secular planting of crops, are

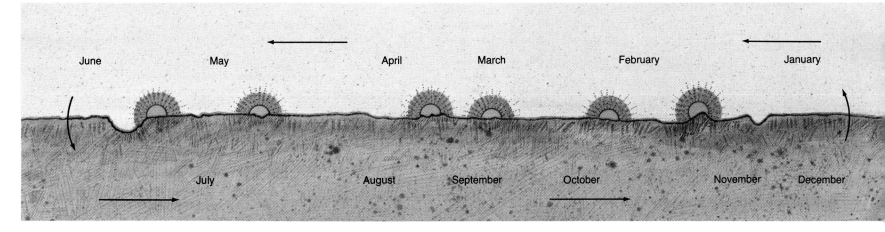

Nor may the Sun Priest err in his watch of Time's flight; for many are the houses in Zuni with scores on their walls or ancient plates imbedded therein, while opposite, a convenient window or small porthole lets in the light of the rising sun, which shines but two mornings in three hundred and sixty-five on the same place. Wonderfully reliable and ingenious are these rude systems of orientation, by which the religion, the labors, and even the pastimes of the Zunis are regulated.

According to Zuni historian Andrew Napatcha, the last Zuni sun watcher, who died in 1953, plotted the seasons on a large, flat stone. The religious leader who assumed some of the sun watcher's duties, Napatcha says wistfully, now uses a calendar.

Among the Hopi the ceremonial and agricultural cycles are closely related, and ceremonial

toward the left—out of winter into spring (January). The sun at the far left has reached the summer solstice and now moves back toward winter. Second sun from left announces the early corn harvest (mid July). Third sun site announces the time for spring planting (March and April). Fourth— start of the main harvest: and fifth, the end of harvest (late October).

determined by solar observations: the *Wuwuchim, Soyal,* and *Niman Kachina.* At First Mesa, the easternmost of three mesas the Hopi occupy in northern Arizona, the observations have traditionally been made from the roof top of the matriarchal house of the Bear Clan, near the center of the main group of houses of Walpi. Walpi is dramatically situated at the westernmost point of the mesa, high above the surrounding plain, with clear views in all directions. Sunrise observations were made from a point south of all the houses in the village.

The Wuwuchim ceremony in November begins the ceremonial cycle. A new fire is kindled and young men are initiated into full adult status in the tribe, both acts symbolizing new beginnings. Next comes the Soyal, a complex nine-day winter solstice ceremony. One of its

171

Above: Huichole girl paints sun symbols on her face as part of her people's annual squash harvest celebration. Daggers of light, above right, pierce sun circles at Colorado's Hovenweep National Monument. Only at the midsummer season do rays reach deep enough into this natural stone corridor to touch the petroglyphs. Right: sun chasers appear frozen in time at Arizona's Canyon del Muerto, a prehistoric burying ground. Sun related rock art appears throughout the southwestern United States.

main purposes traditionally was to turn back the sun from its southward course by depositing prayer sticks at a "Sun Altar" far across the valley in the direction of midwinter sunrise. According to a present-day Hopi tribal leader, the ceremony's modern purpose is to celebrate the renewal of life that the beginning of the sun's northward journey in the sky represents, rather than to express a primitive fear that the sun's southward course would not be reversed. There is also a summer solstice ceremony to celebrate the completion and continuation of this sun-caused cycle of life.

One of the first careful observers of Hopi culture was Alexander M. Stephen, who lived at Walpi from 1891 until his death in 1894. He documented the criteria used by Hopi sun watchers to determine the beginning of Soyal, reckoned as four days after the sun sets in a notch in the San Francisco Mountains known as Luhavwu Chochomo. Stephen's measurements allow this notch to be readily identified as Schultz Pass, which is north of Flagstaff, Arizona, and 84 miles from Walpi. Just to the right of the pass are three small volcanic cinder cones, used as markers, depending on the village from which the solar observations were made.

Similar observations are to this day carried out not only at Walpi but also at the Hopi villages of Shongopovi and Mishongnovi on Second Mesa, says the Hopi leader.

Still another sun-watch procedure that was in use at the Hopi village of Oraibi well into this century involved a small recessed chamber on an interior west-facing wall. The time of solstice was when sunlight entering the room from an opening in the outer west wall illuminated this chamber.

Through a good part of this century the chief sun watcher at the Hopi village of Oraibi has been Don Talayesva. In cooperation with Yale anthropologist Leo W. Simmons, Talayesva wrote a remarkably warm and candid autobiography, *Sun Chief,* published in 1942. One passage tells of his learning that he would be the next Sun Chief:

During these sorrows my uncle Talasquaptewa, who was very old and feeble, said to me: "My nephew, we have looked into your heart and chosen you to be the Sun Chief (Tawamongwi) instead of your brother Ira who is older. You will succeed me shortly in this office. I want you to watch me closely when I make the offerings so that you will know how to do it when I pass away." That meant that I was following in the line of succession—like a king—and would be Sun Chief of Moenkopi as well as Oraibi. He showed me the special place to stand or sit in guiding the rising sun in its journey to its summer house, and taught me all that I should know about the special office.

Writing about his first Soyal, Talayesva described making *pahos,* prayer offerings made of feathers, native string, herbs, and willow sticks. "I learned that this is the most important work in the world, that the gods and the spirits are holding out their hands for pahos, and that if the Soyal should fail, life for the Hopi might end." He describes one of the key rituals in which the Star Priest, representing the sun god, receives a rawhide sun symbol fastened to a stick. Taking the stick in both hands, he shakes it, dances north of the fireplace, and "sideways from east to west and back, twirling the sun symbol rapidly, clockwise," acting out the going and coming of the sun. "The beating of the drum became louder and louder, as the Sun god danced and leaped about in a marvelous manner."

Late in the night kachinas in yellow masks and red horsehair came from the Ahl kiva and danced with the Soyal participants. Then, says Talayesva:

A priest was sent with offerings to Sun shrine (Tawaki) on top of a high mesa about three miles southeast of Oraibi. It was to be a hard journey for him, because he had to run fast, present the offering just as the Sun god peeped over the horizon, and return swiftly. This prayer offering was very important because the Sun god is chief over all, and gives heat and light, without which there would be no life.

Although the Sun Chief's autobiography is completely open about events of his personal life, Talayesva always is careful to emphasize that he would not reveal certain sacred aspects of the Wuwuchim and Soyal ceremonies. "What I did in the Soyal is secret," he tells his editor-collaborator at one point, disclosing no more of his sun-watching procedures.

Recent discoveries have shown that a host of early peoples in North America devoted a great deal of time and ingenuity to sun watching. More than 40 California tribes, for instance, are now known to have recognized and observed the solstices. Ten archaeological sites found in California seem to mark these extreme seasonal travels of the sun. Typical is a stone circle with a stone bisecting line on Cowles Mountain, within the city limits of San Diego. From this point, the first gleam of winter sunrise is bisected by a small rocky peak 16 miles away.

At an Indian stockade of the Fort Ancient culture along the Great Miami River in Ohio, an unusual pattern of posts marks the excavated center plaza. Investigators from the Dayton Museum of Natural History reported last year that alignments between the plaza posts and various dwellings and poles on the periphery appear to define the solstices, the equinoxes, and the May corn-planting time.

At Cahokia, a major Indian urban complex on the Mississippi River near St. Louis dating from the 10th to 14th centuries, a series of pits that once held wooden posts lines up with the solstices and equinoxes. This and the discovery of several interesting artifacts provide evidence that the Cahokians used the sites not only as solar observatories but also as stages for winter solstice ceremonies.

Several of the dozens of large stone "medicine wheels" left by Indians along the eastern flanks of the northern Rocky Mountains are aligned with important astronomical points,

Petroglyph near Santa Fe, New Mexico, bears shield thought to be the sun, with a grouping of bright stars at its center.

including the summer solstice sunrises and sunsets. For instance, the Bighorn Medicine Wheel in northern Wyoming is a rock structure laid out on the ground across a diameter of about 90 feet, with a large stone cairn in the center and six others around the circumference. One of these cairns lies distinctly beyond the edge of the wheel to the southwest, and a line from it through the central cairn points to within 0.2 degrees of the place where the sun rises at summer solstice. A similar line from another cairn on the southeast edge seems to point to sunset at summer solstice. Another medicine wheel on Moose Mountain in Saskatchewan shows similar alignments. But whereas the Bighorn site seems to have been used sometime in the period 1400 to 1700 A.D., astronomical and archaeological evidence indicate the Moose Mountain Medicine Wheel may have been employed nearly 2000 years ago.

At Chaco Canyon in northwestern New Mexico, where a well-developed Anasazi Indian culture thrived from the 10th to 13th centuries, a number of probable solar observation sites have been identified. At one, for instance, near the top of an ancient rock stairway and near a sun glyph painted on the rock wall, an observer would see the sun at winter solstice sunrise almost exactly framed by a narrow rock chimney some 1500 feet off to the southeast. This arrangement may well have been a very accurate device for determining the solstice date.

A dramatic new kind of solar observatory, a type never before seen, was discovered in Chaco Canyon in 1977 and documented in research in 1978 and 1979. The structure lies atop 430-foot-high Fajada Butte at the south opening of the canyon. On a southeast-facing ledge near the top of the butte, three sandstone slabs, each taller than a man and weighing

Left: sand painting shows a kachina spirit messenger at work in the small, sunny world of a Southwest Indian homestead. The kachina opposite represents Tawa, the sun spirit itself. Burst of feathers around its face suggests rays of light. Below: dawn of the summer solstice cuts far cliffs and lines up with a major axis of Wyoming's Bighorn Medicine Wheel. More than 80 feet across, with 28 radiating spokes, the Wheel was probably used to make both solar and stellar alignments for religious purposes.

several tons, lean against the cliff face. Behind and underneath them, two spiral petroglyphs have been carved, one large and one small.

Painstaking photographic observation over many months shows what happens there. Near midday a small amount of light from the sun passes down through the opening between the center and right-hand slabs to form a narrow beam of light on the cliff face behind. During a period of about 20 minutes this beam travels vertically downward through the large spiral, and at summer solstice it exactly bisects the spiral. Through the succeeding months the midday light beam slowly moves to the right, and by winter solstice it is exactly tangent to the right edge of the spiral. By that time there is a second beam of light, created by sunlight passing through the opening between the center and left-hand slabs, exactly tangent to the left edge of the spiral. The effect is dramatic, as though the two parallel beams of light are embracing the spiral. At the equinoxes this second beam of light, now much smaller, bisects the second smaller spiral petroglyph.

So here, atop a wind-swept butte in a now deserted canyon, we have an elegant, complex solar calendar, clearly delineating the beginning of each of the four quarters of the year and, by easy interpolation, the progression of time in between.

I can easily see Chacoan sun watchers, profiled against the sky of Fajada Butte, making their midday solar observations or waiting beneath a sun glyph in the predawn chill for the rising sun to be bisected by a distant outcrop. I can see the Cahokians sighting the rising sun along their array of posts through the morning Mississippi mist. I can see the Indians of the northern Rockies checking their medicine wheel rock patterns for the time of the next gathering of their scattered peoples. I can see early peoples there and at sites everywhere across the continent watching and recording the sun's seasonal progress in ways as numerous and diverse as their ample skills and intelligence could devise.

There is a clear continuum from past to present in the cultural importance of the sun to traditional peoples. I think it probable that all early societies everywhere closely observed the sun and incorporated its movements into their thoughts and cultures. Far more than we do today, our predecessors lived in harmony with the cycles of the sky, and the sun's influence was known to be inseparable from life itself.

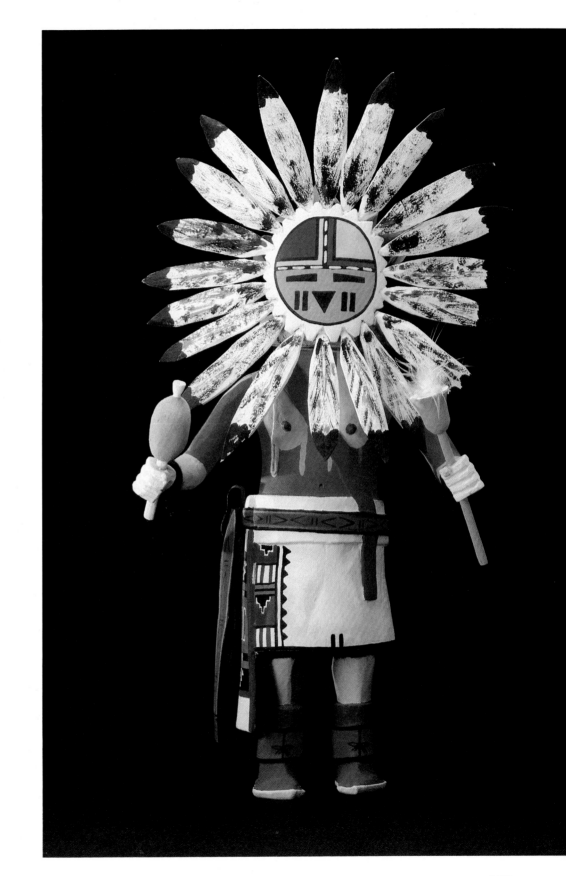

Silvio A. Bedini

Marking Time

A Dial is the Visible Map of Time, till Whose Invention 'twas follie in the Sun to play with a Shadow. It is the Anatomie of the Day and a Scale of Miles for the Jornie of the Sun. It is the silent Voice of Time and without it the Day were dumbe . . . .

So wrote Robert Hegge in 1630. And indeed the moving shadow of the sundial's *gnomon* (from the Greek meaning "interpreter" or "the one who knows") has measured the sun's daily journey and divided the daylight hours for every civilization. The gnomon assumed many shapes, from the primitive denuded tree trunk,

the erect pole in an open area, to the megalith in Majorca (left), or the metal stylus of a fire-gilt monumental dial from a Chinese palace or temple of the Ming period.

As the sundial persisted through the centuries, it underwent constant refinement, proliferating in a multitude of forms, while geared time-reckoning devices such as the clock evolved independently out of a tradition of astronomical models. The earliest surviving artifact of that tradition is a fragment of a Greek geared calendrical or computing instrument from about 87 B.C. Conceivably the Greek device provides the link between the mechanical planetary globe of Archimedes (3rd century B.C.) through the 13th-century Islamic geared astrolabes to the splendid 14th-century mechanical planetaria of Richard of Wallingford and Giovanni de' Dondi, from which the mechanical clock "degenerated." Of these elaborate astronomical models, the mere telling of everyday hours was but a minor function.

The earliest surviving sophisticated time-reckoner, and a predecessor of the sundial, is a shadow clock recovered from an Egyptian burial of the 15th century B.C., although time-reckoning devices may have been known even earlier in Egypt. It consisted of a slate block marked along its length with five equal divisions. At one end a raised projection cast a shadow upon the scale when the block was placed upon a flat surface, with the projection at the east during the morning hours. The shadow crept along the scale, shrinking to zero at noon. The block was then reversed so that the shadow lengthened along the scale from noon to sunset.

A single sundial is referred to in the Bible, in enigmatic texts that have puzzled scholars for centuries. The Second Book of Kings of the Old Testament (20:9–11) relates the story of Achaz, who ruled as eleventh king of Judah, and of his son, Hezekiah. In about 771 B.C. a sundial was made for Achaz by Uriah, the priest of Jerusalem. As a consequence of having sought and obtained a military alliance with the King of Assyria, Achaz had been required to not only pay heavy tribute, but to adopt the religion of the Babylonians, less than two decades after the Babylonian calendar had been rectified. Knowledge of sundials may have come to Judah from that source.

The dial of Achaz was not mentioned again until the thirteenth year of the reign of his son, Hezekiah, who had become mortally ill. Isaiah, the son of the prophet Amos, warned him that

his death was imminent. Hezekiah prayed to God for an extension of his life, which was given, and then Hezekiah demanded a sign in confirmation thereof:

And Isaiah said to him: this shall be the sign from the Lord, that the Lord will do the word which he hath spoken: Wilt thou that the shadow go forward ten lines, or that it go back so many degrees? And Hezekiah said, It is an easy matter for the shadow to go forward ten lines; and I do not desire that this be done: but let it return back ten degrees. And Isaiah the prophet called upon the Lord: and he brought back the shadow ten degrees backward by the lines, by which it had already gone down in the dial of Achaz.

Another reference to the dial occurred in the Book of Isaiah (38:8) with the words:

Behold, I will bring back again the shadow of the lines, by which it is now gone down in the sun dial of Achaz with the sun, ten lines backwards. And the sun returned ten lines by the degrees by which it is gone down.

Opposite, like gnomons of giants' sundials, Minorcan megaliths stand silhouetted against the setting sun. Below, part of the "Antikythera device," a geared Greek calendrical instrument of about 87 B.C. Bottom, the Dial of Achaz, a sundial made by Christoph Schissler in 1578 as an interpretation of an Old Testament text.

177

These passages lent themselves to various interpretations, but it remained for a 16th-century German maker of mathematical instruments to render a practical solution. In 1578 Christoph Schissler of Augsburg constructed a refractive dial based on the Biblical passages. It consisted of a basin engraved with degrees on its inner side, and two gnomons, one in the form of a standing figure of an astrologer holding a staff and a string from the end of the staff, and a pointed rod in the center of the dial. When the basin was filled with water, the time was indicated on the scale by the shadow of one of the gnomons, which the refraction of light caused to turn backward 10 degrees. It is a matter of interest that Schissler produced the dial more than half a century before the appearance of the first published description of the phenomena of refraction.

Near the end of the 17th century, the Schissler dial came into the possession of a member of the Protestant religious group called the Pietists. Fleeing persecution, some Pietists settled in Pennsylvania, where the dial was used for many years to calculate nativities and to determine auspicious days.

Tradition attributes the earliest Greek knowledge of sundials to Anaximander of Miletus of the sixth century B.C., an attribution partially reinforced by the statement of Herodotus in the fifth century B.C. that the concept of the sundial came from the Babylonians. However, the gnomon had been already known to the Greeks from an earlier period, during which it had been used for solar and celestial observations.

One of the earliest forms of Greek sundial was the *hemisphericum*, a block of stone having a hemispherical hollow cut into its horizontal upper surface. A modification of this form was an invention or improvement attributed to Berossos, a Chaldean priest and astronomer living about 300 B.C. He removed the unused part of the hemisphere below the arc of the summer solstice so that the stone was cut according to the latitude in which it was to be used, with the gnomon attached horizontally. This new form was known as a *hemicycle*. By cutting the concavity into the shape of a cone, a greater opening was made possible, permitting more accurate graduations to be inscribed. This version was called a *scaphe* dial (from the Greek word meaning "basin" or "trough").

Another form of the sundial developed by the Greeks in the second century B.C. was the *pelekinon* (from the Greek meaning "double-

bitted axe"). It was adaptable to horizontal and vertical surfaces, with a gnomon at the center. The upper and lower edges were hyperbolas marking the extreme daily paths of the shadow of the winter and summer solstices, while at the equinoxes the path became a straight line. All of the shadow paths were calibrated in seasonal hours.

Knowledge of the sundial came to the Romans from the Greeks, their neighbors and predecessors, by about 290 B.C. By the second century B.C. sundials had become prevalent in Roman communities, as suggested in the works of the Roman playwright and poet Titus Maccius Plautus:

The gods confound the one who first found out
How to distinguish hours! Confound him, too,
Who in this place a sundial did erect,
To cut and hack my days so wretchedly....

A collection of Oriental sundials includes, left to right, a 19th-century Chinese geomancer's or diviner's dial with compass, adjustable sundial ring, and folding gnomon; a "pocket book" sundial from 19th-century China with compass and string gnomon; an 18th-century Japanese pocket "compendium" of instruments

including a magnifying glass, a dark glass for viewing the sun, a compass, and a scaphe (bowl-shaped) sundial; a tiny Japanese pocket compass sundial from the early 19th century; another geomancer's compass dial from 19th-century China marked as a perpetual calendar; and a 17th-century Korean scaphe sundial of brass.

Generally, though, the dials produced in the period before Christ were astronomical models for demonstrating the annual cycle of the sun, while those produced in the period after Christ became diurnal timekeeping devices.

From the fall of the Roman empire, with the interruption of the Byzantine civilization, the line of evolution was resumed in Islam and transmitted to western Europe in the Middle Ages, by means of translations of Arabic texts. The tradition of the sundial continued through the Renaissance and the Reformation with increasing use, culminating in a proliferation of forms, often for scientific purposes.

Much of the Greek and Roman culture continued to flourish in the East for a time, but with the closing of schools as a consequence of political and religious problems, scholars fled to Persia and Syria taking with them Hellenistic learning. The sundial was adopted by the Mos-

lems for their own use in both horizontal and scaphe forms, and it served not only for reckoning time but also to assist the faithful in obeying the imperative of Mahomet for the five daily prayers, during which the believer is required to face in the direction of Mecca. A sundial fixed in the courtyard of a mosque could, when the sun shone, indicate the times for prayer.

A portable compass sundial called a *quibla* was also developed in Islam. It consisted of a circular brass box with a hinged lid containing a compass and a cutaway plate engraved with a semicircular scale of degrees. Numerous Islamic writings on sundials were produced between the 9th and the 15th centuries, of which the 13th-century treatise by Abu 'Ali al-Hassan (b. Ali b. Umar al-Marrakushi) of Morocco was the most comprehensive. The earliest known surviving Islamic sundial was made in 1159–60 for the Zangid sultan of Syria.

The sundial was known also in China in several forms from at least the first century B.C. A stone equatorial dial of the third century B.C., recovered by archeologists in 1932, was marked to divide the day not into standard hours but into 100 equal parts, each equivalent to 14 minutes in length. Then the art of the sundial appears to have been lost in China for a considerable period and rediscovered in about the 12th century.

One type prevalent in China was a monumental sundial placed in the equatorial plane to receive the sun's rays on the dial plate's northern face during the first half of the year and on its southern face during the remainder. In other words, the gnomon's shadow was cast on the underside of the dial plate in the seasons during which the sun is below the equator. Inscribed on the upper and lower faces were the signs of

Below, a French cannon sundial of the late 18th or early 19th century, signed "Rousseau Paris." Right, top to bottom, a French ring dial, dated 1640, with a replica of a 10th-century Saxon pocket dial found at Canterbury Cathedral; a late 18th-century Persian quibla, or portable dial for determining the times of the five daily Islamic prayers; an 18th-century French "Butterfield" pocket sundial with compass and folding gnomon; and a German pocket diptych sundial of ivory dated 1518.

the zodiac, the shadow of the gnomon remaining in each sign for a period equal to two Western hours. However, the sun's shadow would not appear on either face during the equinoxes, as the dial was parallel to the sun's rays at those times. Accordingly, both faces of the dial were graduated with the zodiacal or hour symbols, and a long gnomon of metal was inserted through the dial's center in the polar axis and extended beyond each face to a length approximately equal to the radius of the dial plate.

Sundials were also in common use in Japan and Korea from early times, derived from China's Han culture. Sundials used as public timekeepers were erected during the Sejong reign in Korea, and in time various types were imported in great numbers from China. The Chinese dials were copied in both Japan and Korea with some modifications, as were dials later introduced from Europe.

Invasions and wars which ravaged the Old World between the fourth and sixth centuries brought such havoc and economic decline that concern with timetelling diminished considerably. Throughout this period, however, the Christian church continued to use sundials to establish the hours for prayer, and subsequently also played a major role in the development of monastic alarms operated by water clocks, predecessors of the mechanical escapement clock.

An early form of sundial used by the Church was the "scratch dial," a type peculiar to England as early as the seventh century. Relatively primitive in form at first, it consisted of a semicircle with the hour lines radiating from the center where a horizontal gnomon was inserted. At first these dials were literally scratched or inscribed directly into the surface of the stonework on the southern walls of churches. Later they were produced as separate stone slabs which were attached to the walls.

Portable dials were also used during the same period. A remarkable survivor is a Saxon pocket dial of the 10th century found in the soil of Cloister Garth of Canterbury Cathedral in 1939. It is in the form of a slender silver tablet with a gold cap and chain to which a pin is attached to serve as a gnomon. It was similar in design to the "shepherd's dials" which became common throughout Europe in a later period.

The mechanical escapement clock emerged in the late 14th century, a product of that late medieval European interest in mechanics which would ultimately give rise to modern

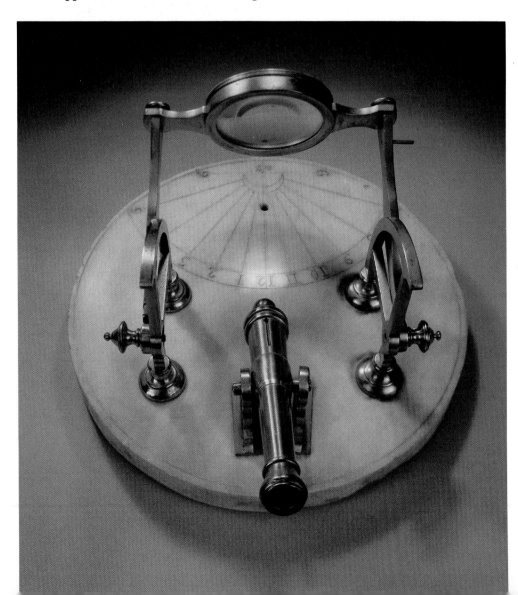

technology. The sundial played no role in the clock's evolution, but continued in use long after it was perfected, following a separate evolutionary path while serving the same common needs. An extremely tenuous relationship between clock and sundial might be assumed through the invention of the toothed gear about the time of Archimedes, possibly for use in models of the calendar. A further relationship might be established with the *clepsydra* or water clock as an antecedent of the anaphoric clock, which was basically a planetary model or star map made to rotate at a rate comparable to the actual rotation of the heavens.

Just how much of this technology was preserved and adopted by the Church is not certain, but it reappeared in the monastic water clock of the Middle Ages. As the Church regularized the monastic day during this period, it also shed its influence upon the community and imposed the same conformity upon society as did Islam with its regular calls to prayer. The records of English monastic houses in particular are rich in references to the construction of *horologia* in the last two decades of the 13th century, indicating that there was considerable activity in the production of timepieces for ecclesiastical use.

In 1309 an iron clock was installed in the tower of the church of San Eustorgio in Milan, and others followed within a few years. Those cited to have been mechanical instead of water clocks were erected at Milan in 1335, Modena in 1343, Padua in 1344, Monza in 1347, and Strasbourg in 1352. By the middle of the 14th century public clocks with bells, jacquemarts, and other automata had become almost commonplace throughout most of Europe.

The advent of the mechanical clock had little effect on the production of sundials, which continued to be made in great variety and numbers from the 15th century onward. Guilds of craftsmen specializing in the art of dial making were formed in the Low Countries and southern Germany, and some of the sundial makers achieved great prestige with their work. Many of their dials combined the skills of the goldsmith and the engraver and became showpieces in the cabinets of curiosities of princes and prelates.

From time to time new forms of sundials were introduced in these centers. One of the most popular was the equatorial dial, based on the earlier version of the scaphe dial and first described in the first quarter of the 15th century. The gnomon, now called the style or stylus, was set in the direction of the Earth's axis perpendicular to the plane of the celestial and the terrestrial equator. The equatorial dial remained in great demand until the end of the 18th century.

Interest in gnomonics continued during the next several centuries, and the works of more than 500 authors in various languages on the subject have survived. During this period a proliferation of dials in many forms appeared on the walls of churches, public buildings, and the homes of the wealthy. Many featured such refinements as the "Babylonian hours," marking the time from sunrise, and the "Italian hours," marking the time from sunset. The dials frequently also served to indicate the date, for the style described the straight line of the equinoxes and the "arcs of the signs" or hyperbolic curves corresponding to the days—the twenty-first of each month—when the sun entered one of the signs of the zodiac.

An unusual form of sundial was the cannon dial, invented in the late 18th century. Although attributed to a French military engineer named Jacques Rousseau, it was actually first conceived by the Comte d'Angivillers, superintendent of the French royal buildings. He planned to install a small cannon on the terrace of the Damartaine and equip it with a powerful magnifying glass positioned in such a manner that it would ignite the cannon by means of the sun's rays at noon. The project never came to realization, but in 1784 a device of similar conception was installed in the Jardin des Plantes by Comte Georges Louis Leclerc de Buffon, the naturalist. He devised a contraption featuring a magnifying glass which focused the sun's rays upon a thread supporting an arrangement of hammers. When the heat of the sun burned through the thread, the hammers fell and struck 12 times upon a great bronze bell.

In 1789 or 1790, engineer Rousseau installed a small mortar on the terrace near the Café de la Rotonde in the grounds of the Palais Royale, a dining place which was to be remembered in relation to the French Revolution. He attached a strong magnifying glass to the mortar by two adjustable arms, and oriented it to the sun at exactly high noon. The cannon was loaded with black powder, some was also added to its touch hole, and at noon the sun's rays ignited the powder and fired the cannon.

The cannon dial became instantly popular as a place for ladies and gentlemen to gather each

Below, a German armillary sphere with clockworks, made by Josias Habrecht of Strasbourg in 1572, exemplifies the 16th and 17th-century European mechanical masterpieces that briefly overshadowed the clock itself. Right, a page from the mid-15th century French manuscript, The Clock of Wisdom, reveals the period's compelling interest in time measurement. Shown are gravity and spring-driven clocks, an alarm device, an astrolabe, an horary quadrant, and three kinds of sundials.

day, ostensibly to set their watches but also to be seen and to exchange the news and gossip of the day. Rousseau's cannon dial remained in use, with interruptions, until about 1916. Meanwhile small operating models were commercially produced by a number of European makers of mathematical instruments and enjoyed a wide popularity.

The use of sundials continued well into modern times. A writer on dialing in 1719 wrote:

And tho' we be furnished with some kind of moving machines . . . as clocks and watches, yet these are often out of order, apt to stop and go wrong, and therefore require frequently to be regulated by some unerring instrument, as a dial; which being rightly constructed, will always (when the sun shines) tell the truth.

Despite the profusion of sundials being produced and marketed, not everyone could afford to own one. Rural dwellers generally relied on the simple device of the "noon mark," consisting of a groove or furrow cut into the sill of an oft-frequented doorway or window ledge of the home, or the threshold of a barn. The mark was made at the point or line upon which the sun's shadow fell at exactly noon. Country folk became quite adept in determining the relative time by the location on or variation of the sun's shadow from the mark. Those who could afford the modest expenditure for an inexpensive dial often used a window dial. The simplest and most easily produced form of the horizontal sundial was cast in lead or pewter to fit a window ledge and was designed for the particular latitude in which it was to be used. English colonists in North America commonly used noon marks or window dials during the first century of settlement, during which very few mechanical clocks or watches were brought to the colonies.

By the 19th century the technological refinement of the clock had reached a high degree of precision, and mass marketing methods made even complicated timepieces available to the public at modest prices. Consequently the sundial fell into disuse, and was gradually produced only in decorative forms relegated to the enhancement of public parks and family gardens.

Nonetheless, new forms were occasionally developed and revived, even to the present time. Most popular of these is the heliochronometer, an ornamental as well as functional form of the equatorial dial. It is equipped with a gnomon aligned with the polar axis of the Earth, is automatically compensated for the equation of time, and is calibrated to be read in minutes. The heliochronometer was first introduced in the late 19th century and installed in railroad stations in France and Switzerland to provide a comparatively accurate time reckoner for travelers to use in setting their watches. It was "reinvented" and patented in 1912 by William Renard Pilkington of Lancashire, England, and commercially produced in substantial numbers. More recently it has been revived in the United States.

Meanwhile, the multitude of ingenious and artistic products of the sundial makers through the centuries now serve little purpose and have become the prized possessions of collectors and museums.

Day and the Hours, a decorative sundial designed in 1916 by American artist Paul Manship, reveals the inspiration he found in the art of India.

For millennia artists have created images and symbols for the sun and light. But as post-Renaissance man turned his attention increasingly to the analysis of the world around him, the sun gradually lost its ancient value as a religious and astrological symbol and gained new importance as the source of natural light in the visual arts. Baroque artists, for instance, often used light to intensify the drama of their images of people, an approach epitomized by Rembrandt.

The culmination of this interest in the physical properties of light occurred in the naturalis-

The Light of Art

Valerie J. Fletcher

tic painting of Europe and America in the 19th century. Romanticism, dominant in the early 1800s, introduced a new emphasis on empirical naturalism in painting. Preferring landscapes over traditional religious scenes, Romantic artists examined the physical phenomena found in nature, not the least of which was sunlight. But the Romantics' pursuit of realism was not an end in itself; rather, they used it to express their profound spiritual relationship with nature. Thus they were particularly fond of sunrise and sunset, moments rich with color harmonies, emotional and symbolic interpreta-

tions. Sunrise, for example, tended to have optimistic connotations, while the vivid colors of sunset could imply the powerful emotions of anguish or desire. Sunset also suggested the sadness of things coming to an end—death—or even the eternal life after death.

Among the Romantic artists most concerned with the sun's light was the British painter Joseph W.M. Turner (1775–1851). Impelled by his desire to depict such elemental forces of nature as water and air, he devoted years to the scientific analysis of sunlight and its portrayal on canvas, and the light emanating from the

Left, Woman in Morning Sunlight, *1809, by Caspar David Friedrich (1774–1840); from the Museum Folkwang Essen, Essen, Federal Republic of Germany. Below,* Regulus *, 1828–37, by Joseph W.M. Turner (1775–1851); from* The Tate Gallery, London.

October in the Catskills, *1880, by Sanford Gifford (1823–80); from the Los Angeles County Museum of Art, Los Angeles, gift of Mr. and Mrs. Charles C. Shoemaker, Mr. and Mrs. J. Douglas Pardee, and Mr. and Mrs. John McGreevey.*

sun eventually became paramount in his art. Turner studied the esoteric texts of the 16th-century Italian theorist G.B. Lomazzo (who equated sunlight with God), as well as the theories of his contemporaries Henry Richter and Johann von Goethe. Turner himself wrote complicated essays on the nature of light, demonstrating scientifically that "light is color," a belief embraced by many artists who came after him.

While early works reflect the artist's objective analysis of light, Turner's later paintings represent the sun in a more mystical and symbolic fashion, with its light consuming nearly all other elements, including man. Sunlight came to symbolize not only the animation of life, but its torments as well. When the artist declared shortly before his death that "the sun is God," it was a god capable at times of a frightening and destructive power, as in *Regulus*

(1828–37). In this picture Turner referred to a tale from ancient history in which the Carthaginians punished the Roman general Regulus by cutting off his eyelids and facing him to the sun, blinding him slowly. Curiously, as art historian John Gage has explained, none of the tiny figures in the harbor scene actually depicts the condemned man. Rather, in looking at the painting the viewer assumes the role of Regulus being tortured by the searing light.

Reviewers commented on the contrast between Turner's sunlight and the traditionally muted tones of paintings from earlier centuries. One critic noted that in *Regulus* the "sun absolutely dazzles the eyes," while another observed that the painting, like the sun, cannot be looked at from up close; rather one must back up across the room, where still one sees only "a burst of sunlight." Turner's intensely personal attitude toward the sun is echoed in the literature of the period. In his poem, "The Demon of the World," Percy Bysshe Shelley spoke of "the sun's broad orb," its "intolerable radiancy," and how in contemplating a fiery sunset "one's rapt imagination soared."

In contrast to Turner's tumultuous scenes of vertiginous sunlight, the German Romantic artists preferred silent landscape vistas in which light subtly pervades vast spaces. Caspar David Friedrich (1774–1840) and his colleagues wished to express their transcendental experience of nature, an artistic goal stemming from the pantheistic doctrine that spread in Germany during the early 19th century. Abandoning traditional church-oriented worship, the German pantheists believed that God manifested himself in the various forces of nature. Communion with nature became their means of communication with God, and their paintings sought to visualize this emotion, often by means of radiant light. Recalling perhaps the Biblical injunction that "God is light" (John 1:5), Friedrich compared the "radiating beams of light" in one of his paintings to "the image of the eternal life-giving Father." In *Woman in Morning Sunlight* (1809), the glowing sunrise can be construed as a symbol of divine presence in the world or, more broadly, as an image of spiritual succor, or as a presage of eternal life. Although the painting shows the seemingly commonplace event of a woman watching the dawn, the emergent sunlight imposes both silence and quiescence, transforming this image into a picture not unlike the altarpieces in medieval cathedrals, in effect an icon for

spiritual meditation.

A comparable feeling pervades the landscapes of the American Luminists of the mid-19th century, as seen in Sanford Gifford's *October in the Catskills* (1880). The American attitudes toward light and nature were remarkably similar to those of the German Romantics, both of which were derived from some of the same sources. Scenes of vast mountains and prairies, suffused with light, expressed a subliminal optimism in America. Radiance symbolized confidence in the country's future, including the political doctrine of Manifest Destiny, and expressed the prevalent belief that America was particularly blessed by God. The spiritual overtones of sunlight in Luminist art corresponded to the transcendentalist philosophies widespread among writers and artists. Preeminent among these was Ralph Waldo Emerson, whose frequent reference to light in his writings included the thought that pure light was "the reappearance of the original soul."

Le Moulin de la Galette, Montmartre, 1876, *by Pierre-Auguste Renoir (1841–1919); from the Louvre, Paris.*

While the Romantics imbued their sunlit images with spiritual qualities, the Impressionists of the 1860s–90s devoted their art to sunlight's purely visual aspects. Rejecting religious and poetic interpretations in favor of greater objectivity, the Impressionists painted few emotionally significant sunsets. They also avoided depictions of the sun's disk itself, with its inherent and traditional symbolisms.

By going outdoors to paint—a rare practice among earlier artists—the French Impressionists immersed themselves in the light that they so rigorously analyzed. Influenced in part by scientific essays on light and color, they considered white sunlight as the sum or composite of the pure colors of the spectrum. Believing that sunlight and its colored reflections reached even into shadows, they banished black from their canvases. Using unadulterated, high-keyed colors, they created landscapes of unprecedented luminosity.

By 1876 critic Edmond Duranty noted the Impressionists' "intoxication with light" and admired the paintings in which the viewer can feel "the light and heat vibrate and palpitate." In 1886 Felix Feneon wrote that "sunlight was at last captured on their canvases." When Pierre-Auguste Renoir (1841–1919) exhibited his masterpiece *Le Moulin de la Galette, Montmartre* (1876), critics recognized that "the effect of strong sunlight falling through foliage on the figures" was indeed the artist's primary concern and that he used "a happy light reflected in the shadows" to create a light-hearted mood. In this picture Renoir culminated his study of sunlight, allowing it to unify the overall composition, soften forms, and dissolve details.

Of all the Impressionists, Claude Monet (1840–1926) was most obsessively concerned with the varying qualities of sunlight and its effects on the perception of objects. In the early 1890s he devoted himself to two series of paintings that represented the visual properties of sunlight with unprecedented attention to detail. In both the *Haystack* and the *Rouen Cathedral* paintings, Monet systematically scrutinized the subtle, nearly imperceptible changes in sunlight from dawn to dusk through the various seasons.

Monet discovered that he required a whole sequence of paintings to record the nuances of ever-changing light. He worked on each painting for no more than half an hour per day, but even so found it difficult to capture and define

the fleeting changes, writing in despair that "the sun goes down so quickly that I cannot follow it . . . ." Viewing the twenty *Rouen Cathedral* paintings in one gallery in 1895 was like watching the sunlight travel across the church's facade from early light to sunset, all within a short span of time. Since the image itself remains constant in all the pictures, the real subject of these series is light itself. This represents the apogee of the 19th-century interest in observing and recording the physical properties of sunlight.

As the Impressionists exhausted the purely visual possibilities of sunlight, a general reaction set in against such single-minded empiricism. Post-Impressionist artists either sought a more theoretical foundation for their paintings or they gave freer rein to personal expressiveness, often in symbols.

The Neo-Impressionists, led by Georges Seurat (French, 1859–91), elaborated a complex theory of complementary colors that gave them a more accurate depiction of sunlight. Believing the paintings of the Impressionists to be too casual and informal in their study of light as color, Seurat built his theories on the precise laws of optics, particularly those established by

the experiments of physicists Ogden Rood and Hermann von Helmholtz. Thus, although Seurat's masterpiece *Sunday Afternoon on the Island of La Grande Jatte* (1884–86) was painted entirely in the artist's studio, it effectively simulates vivid sunshine at four o'clock in the afternoon. Seurat scattered tiny dots of a high-keyed "solar orange" among the other colors to create the illusion of sunshine, with more orange dots in the "sunlit" areas and fewer in the shade. A young painter, Maurice Denis, in his 1906 essay *The Sun*, described Seurat's depiction as "the cool sun . . . with the charm and delicacy of incomparable nuances, harmonious and composed of perfect rhythms in precise equilibrium."

Seurat called his style of painting "Chromo-Luminism" to emphasize his commitment to colored sunlight above all other artistic concerns. Yet, by treating sunlight conceptually, entirely by means of color theory, Seurat had also moved the light a step away from the artist—sunlight in art was no longer to be based primarily on direct observation. This translation of sunlight into scientifically determined combinations of pure colors was to culminate in the paintings of Robert Delaunay.

The Sun, 1909–16, by Edvard Munch (1863–1944); from the Festival Hall, University of Oslo, Oslo.

To some Post-Impressionist artists, the sun became a profound personal symbol whose light could be used to express emotions, a purpose diametrically opposed to that of Seurat. One of these was the Norwegian Edvard Munch (1863–1944). He often included the midnight sun in his early expressionist landscapes, where it and its reflection on the sea served as an unprecedented symbol of sexual desire. But in a cycle of 11 murals for the University of Oslo, Munch perceived the sun as a more transcendent and universal image. *The Sun* (1909–16), with its centralized, iconic composition, radiates an intense, stylized light that invites metaphysical interpretation as the origin of man's spiritual and intellectual existence, as well as the source of all physical life. In illuminating the world and its contents, the sun's light excites man's desire for understanding and stimulates his creativity in all fields, including the arts and sciences. In addition, the sun symbolizes the incomprehensibility of the universe and of such concepts as eternity and infinity.

The writings of Munch's contemporaries also incorporate philosophic symbolism of the kind expressed in *The Sun*. August Strindberg published a book entitled *Towards the Sun*, which Munch illustrated. Likewise in his grandiose metaphysical essays, such as *Also Sprach Zarathustra*, Friedrich Nietzsche often used cosmic or primordial light as a central theme. Munch himself believed that life was a cosmological power inextricably allied with the sun's light, and he wrote that his art aspired toward "the heavens of the sun's realm."

Of all the Post-Impressionists, however, the Dutch painter Vincent Van Gogh (1853–90)

was the most concerned with the sun, not only as a source of light and color but as a symbol significant to him personally. He accorded it a special place in his art, situating it prominently in the sky of several landscapes, as in *Wheatfield with Reaper* (1889). Here the disk of intense golden yellow dominates the high horizon, a central iconic position recalling the suns in Friedrich's works. Like the Romantics, Van Gogh had moved away from traditionally defined Christianity, yet retained an indomitable belief in and aspiration toward an omnipotent deity. His suns blaze forth a supernatural radiance, implying a divine presence yet surpassing any simplistic equation with the Christian God. Rather they suggest a symbol akin to Turner's and evoke an older relationship

between man and sun, reminiscent of ancient cultures.

For Van Gogh the sun was the central symbol of his quasi-mystical relationship with nature. It represented for him the source of life and well-being, including the physical and mental health that eluded him. During his stay in the mental hospital in 1889, the sun recurs in both his art and his letters: "through the barred window I see . . . the rise of the sun in its glory." In addition to the sun's disk itself, Van Gogh attributed a strongly optimistic significance to its golden yellow color, which appears in many of his paintings.

While living in his sunlit yellow house in southern France, he painted a series of pictures of golden sunflowers, as if to bring the curative powers of the sun into his rooms. In Robert Rosenblum's words, Van Gogh conceived of his sunflowers "as earthbound metaphors for the miracle of the sun's energy." And as Reinhold Hohl has pointed out, Van Gogh even used drugs as stimulants in his keenly felt quest for "the most vital yellow" of sunlight.

Other 20th-century artists used the sun for its purely formal properties. Elaborating upon Seurat's theory equating sunlight with the elemental colors, Robert Delaunay (French, 1885–1941) removed his color theories even further from their source in the sun than Seurat had. In his "Circular Form" paintings such as *Simultaneous Contrasts: Sun and Moon* (1912–13), Delaunay's colors bear no reference to the sun's light in relation to the earthly world.

Anticipating this new role in 20th-century aesthetics, Delaunay used the sun as an element of formal composition. And today, as both personal symbol and abstract sign, the sun appears in the works of many artists, including Paul Klee, Wassily Kandinsky, Joan Miro, Isamu Noguchi, and Richard Lippold.

Edwin Kiester, Jr.

The Light and the Dark

The men came tumbling out into the Antarctic snow, chattering with anticipation and bundling their heavy winter clothing around them. The buildings of the U.S. research station were still shrouded in gloom. The clock said it was almost midday, but most of the scientific and naval personnel who were "wintering over" on the frozen continent had given up wristwatches months before. The sun had disappeared last April and now it was September. Night and day had been indistinguishable for nearly five months.

But this very day, in a few minutes, the sun was scheduled to return. The men kept their eyes trained on the jagged ice peaks on the far horizon. They fell silent as a faint, yellow-white tinged the sky; then they began to buzz excitedly as it widened and brightened. Years later, one of them was to remember the next minutes in vivid detail. "There was a kind of green flash. Then inch by inch this thin yellow arc, like the rim of a fingernail really, pushed itself just above the horizon."

A great shout went up from the group. The men began to pound each other on the back.

Mad dogs and Englishmen....

They cheered and applauded. It seemed that all of them had brought cameras and as one they trained them on the pallid disk to record the moment of reappearance.

"It was like a football game," the witness laughs ruefully now. "Grown men, dignified, serious scientists were yelling and congratulating each other as if their team had just scored a touchdown. Some of them hadn't been able to sleep for two or three days because they were so excited about the sun coming back. Of course, the sun set again in a few minutes. We drank a toast to the sun and went back indoors. The darkness settled in again, but now it didn't matter. The sun had come back and everybody felt better."

For most of us the coming and going of the sun is seldom so dramatic or so celebrated. As with Tevya in *Fiddler on the Roof*, "Sunrise, Sunset" bounds our daily routine, unremarkable for its predictability and regularity. Yet we also know that light is life, and the presence or absence of sunshine can have profound effects on us—emotional, physical, cultural, psychological. We know, too, that the sun has shaped man's history and customs in ways that still survive. The very words we use show how the sun insinuates itself into every aspect of living. We describe a cheerful person as having a "sunny" disposition. An episode of sadness is a "dark" day.

Primitive man was confronted by two immutable facts: the sun above and the earth below. He well may have ascribed supernatural qualities to both. It was apparent to the early humans that the sun bestowed the gift of life and the earth nurtured it. He named them Father Sun and Mother Earth, gave them personalities, and made up tales about them.

The San of the Kalahari Desert in Namibia (South West Africa), nomadic hunters who live much as their forefathers did, still tell a charming tale of the sun's origins. The sun, it seems, was originally an ordinary mortal who lived among their tribe but was not one of them. He was an old man who lay on the ground in his hut, and when he raised his arms, light shining from his armpits illuminated the ground around him. But the beams were limited to his immediate vicinity. To spread this light around, an old woman suggested that the children of the tribe steal up on him while he was sleeping, lift him by the arms and legs and throw him into the sky. Time after time the children would creep toward the sleeping man, but each time they approached he stirred and his light shone on them, driving them back. At last, after many tries, they succeeded, lifted him all together and hurled him into the sky. "Stand fast!" the tribesmen cried. "Thou must stand fast, then move through the sky while thou art hot." And the sun has shone on the Kalahari ever since.

Someone who holds the power of life over you is a god, and early man everywhere deified the sun. They gave him different names: the Egyptians worshipped Amun-Re, the Greeks Helios, the Japanese the sun goddess Amaterasu. The Mayas, Toltecs, and Aztecs of Central America had a whole pantheon of sun gods, representing different aspects of the sun. Elaborate rituals were devised to court the sun's favor, for it was quickly realized that the sun was a fickle god, hiding his face when the crops most needed him to smile on them, scorching the earth when the need was for rain. A human sacrifice was not too much to pay for sunshine, and the concept of sacrifice was closely bound up with sun worship. The Aztecs regulary fed the hearts of sacrificial victims to the sun to strengthen him for his daily journey.

Shrugging off all the ritual honor, the sun came and went as he chose, leading to the fear that he might one day disappear completely and leave the world in darkness. A child's night terrors and the Antarctic scientists' depression during the polar night reflect this old indwelling dread. Fear of the dark may have led to the use of fire, a substitute sun which brought light to darkness and quicky developed its own ritualistic aspects—as the symbolic lighting of the torch at the Olympic Games and the use of candles in religious ceremonies still testify.

The ultimate loss of the sun was the eclipse. Early man feared this catastrophe above all others. He invented numerous explanations: the sun had been kidnapped by vicious beasts or demons or even by evil people. He had to be ransomed, either by sacrifice or by the intervention of other gods. These legends still survive in many places. In February 1980, a total eclipse darkened the island of Lamu, off the coast of East Africa. Smithsonian folklorist Peter Seitel questioned the local people about the event. Many of them, he found, accepted the explanation that the Earth was passing through the moon's shadow. Others felt that the sun had lost its way and strayed from its normal orbit. But many told Seitel of an explanation they had learned in childhood:

A giant serpent swallows the sun or moon to cause an eclipse. The round shadow that passes across the shining disk is the mouth of the serpent as it slowly ingests its prey. In the old days, the women of the tribe would beat on logs and perhaps drums and use other noisemakers to frighten off the serpent.

The ritual, of course, was reinforced by the fact that the serpent did give up his quarry.

In almost all cultures and religions, there is a festival that celebrates the return of the sun. In the temperate climates, this holiday often coincides with the winter solstice, the time the short winter days begin to grow longer in prelude to spring, about December 22. The evidence is strong that Christ was born in the spring, but as the new religion spread through the Roman Empire, it adopted some rites from the competing faith of Mithraism.

A spiritual import from the Iranian plateau, Mithraism honored *Sol Invictus*, the Unconquered Sun, and observed a major holiday at the winter solstice. Since both religions celebrated the return of light into a dark world, their respective sacred days quickly merged into one. Even today, the most lavish and joyous Christmas celebrations occur in northern regions where twinkly Christmas trees and blazing Yule logs signal the end of deep winter darkness.

When a temporal leader wanted to proclaim himself divine, he immediately wrapped himself in solar majesty. Louis XIV, the ultimate absolute monarch, assumed the title *Le Roi Soleil*—the Sun King. Hammurabi, the codemaker of Babylon, did likewise, only 2000 years earlier. The great leader of the sun-worshiping Babylonians declared, "I am the Sun of Babylon which causes light to rise over the land of Sumer and Akkad." The alliance between sun and throne still survives in the regal crown. Its circlet of gold with radiant points represents the sunrise and its rays.

Of course, modern worshipers no longer revere the sun—or do they? A visit to a Christian church finds solar symbolism everywhere. In Roman Catholicism, the monstrance, in which the Host is placed for veneration, resembles a sunburst; the priestly robes are decorated with gold in sunlight streaks and shining orbs. The halo above the heads of Christ and the saints represents an ever-shining sun. The Christian Bible is shot through with references in which Christ and light are synonymous. The war between good and evil in the Dead Sea Scrolls is depicted as the battle of the Sons of

Louis XIV, the Sun King.

Light against the Sons of Darkness.

Seeking warmth and light, the American population has divided the nation into a Frost Belt and a Sun Belt. Promoters try to speed the migration by giving their resorts names like Sunny Isles and the Posada del Sol.

The cultural effect of many such efforts has been to turn the human caste system upside down. Once upon a time a man with a bronzed brow and tanned arms labeled himself lower class: he obviously had to toil outdoors in the beating sun. The upper classes were white and determined to stay that way. A real lady of the 1800s cultivated a fashionable pallor. When she ventured outdoors, she protected her pale complexion with a sunbonnet and parasol. If an errant sunbeam did tinge her cheeks, she covered the blush with face powder.

The Industrial Revolution made shade cheap and sun costly. The pale forehead became the mark of the factory laborer and office worker. A "fantastic tan" proved that its owner had both the leisure and the wherewithal to travel to a far-off resort and bask in the sunshine while others were working. To turn up with a golden glow in Boston in February made you the envy of your pale Back Bay neighbors.

When the sun touches the skin, it stimulates melanocyte cells in the epidermal cells to produce a pigment called melanin, which rises to the skin's surface and darkens it. The darkened skin—that "fanstastic tan"—blocks the sun's rays from pentrating the surface and destroying

In all darkness situations studies show people's body clocks run on a cycle of about 25½ hours without sleep.

the underlying tissues. A good tan can block out 95 percent of the damaging rays.

The need for sunlight in limited quantities has colored the face of the world. Were a blond, blue-eyed Scandinavian actually dark he could not capitalize on the few wan rays available in his northern latitudes. Tropical and desert peoples have the opposite problem of excessive sun. Their dark skins block all but the necessary quota of sunlight.

Early man sensed that sunshine brings health, but only recently has science confirmed his observations. Sunlight is essential to calcium metabolism—the process by which we build bones and teeth. When certain invisible rays of sunlight, those of the ultraviolet spectrum, touch the skin, they stimulate production of a hormone we have named Vitamin D. The new substance is taken up by the circulation, and, as it passes through the kidneys, is transformed into an activated form enabling dietary calcium to be absorbed from the intestine. With the aid of parathyroid hormone, the calcium is then synthesized into bone cells. Not much sun is needed for bone building. A New York office worker gets enough just walking to the bank during lunch hour.

A century ago, lack of vitamin D was a major health problem. In the American and European industrial cities, many children in the bone-building years developed a vitamin D deficiency disease called rickets. The disease occurs when not enough calcium can be absorbed for proper bone formation because of underproduction of vitamin D. The long bones of the arms and legs, in particular, soften and weaken, then curve under stress. Knobs form on the ends of the bones at the joints. The cardinal symptom is bowleggedness.

London had more rickets than any other city, and the cause was plain to see. A pall of coal smoke from hundreds of thousands of coal-burning grates as well as from booming factories overhung the city and blocked the sunshine from reaching its residents. Tall buildings and narrow streets combined to provide even greater shadow. Children of the poor suffered most, because they had little escape from their crowded, darkened neighborhoods, and dark-skinned immigrants from India, Pakistan, and other parts of the Empire were the worst afflicted, because their natural pigmentation limited the sun's scant benefits. In the Edwardian era, the problem was so severe that even animals in the London zoo developed rickets.

Rickets today is largely history in American and European cities, because the air has been cleaned up and alternative sources of vitamin D have been found. Cod liver oil won the name of "bottled sunshine" because it was found to contain a form of the vitamin resembling the necessary version; a whole generation developed strong bones while manfully making wry faces and swallowing the bad-tasting stuff.

Ironically, vitamin D deficiency now may be a problem at the other end of the age spectrum. Osteomalacia (softening of the bones) appears to be increasing in older people confined to nursing homes or kept indoors by illness. An English study shows the disease markedly more common in the gloomy English winter than in summer. In the United States, Robert J. Wurtmann of the Massachusetts Institute of Technology has found that residents of a Boston veterans' home, living indoors under fluorescent light, in seven winter weeks absorbed only 40 percent of the calcium needed for proper bone formation. Fortunately, exposure to sunlight or artificial ultraviolet rays could correct the deficiency.

Some of the other healthful effects of the sun are more subtle. Ultraviolet light kills bacteria, which is why rays of an appropriate wavelength are used in operating rooms. But it has also been observed that postsurgical patients heal faster in beds near hospitals windows than in those near interior walls. Arthritics say their condition improves in the sunlight of Arizona, although whether the benefits derive from light or simple warmth would be difficult to prove.

It is a fact, however, that the most successful therapies for the skin condition called psoriasis use ultraviolet wavelengths in combination with a prescribed medication. Ultraviolet is also the treatment for neonatal jaundice and is credited with saving 25,000 premature babies a year. The condition occurs because the immature liver cannot rid itself of a substance called bilirubin, which can then accumulate and cause brain damage. Sunlight is also accepted treatment for herpes infections, and the Russians routinely prescribe sunshine to combat black-lung disease in coal miners.

But the sun is not always medicine's friend; it can also be a deadly enemy. Mythology says that the chariot approached too near the sun and was scorched; anyone who has basked on a beach too long understands the symbolism. Too much sun exposure on unprotected skin brings about a condition called erythema, better

Eskimos on Baffin Island are almost as happy to see the Sun return as are their Huskies.

Depicting Hell as a place where people are condemned to live without light.

known as sunburn. Dermatologists measure it by Minimum Erythemal Dose. One MED is the amount that turns the skin pink and starts melanin production. Five to 10 MED without prior exposure will bring on, four to 14 hours later, the excruciating symptoms of lobster-colored skin that is hot to the touch, acutely painful, and blistered. Even the worst dose, however, usually subsides after about four days, by which time melanin has been mobilized in sufficient quantities to protect against new exposure.

If a few days of sore shoulders were the sole result of too much sunshine, improved appearance might be a fair tradeoff for brief discomfort. Unfortunately, repeated, prolonged exposure damages the skin beyond repair. The constant assault of the sun breaks down the skin's protective features, thinning the epidermal layer and stiffening the underlying fatty tissue. One outcome is the leathery face of the legendary cowboy. Another is the wrinkled furrows we associate with old age, although they are caused by the sun-damaged skin's loss of resiliency, not by old age; septuagenarian and octogenarian blacks and Asians, protected by melanin, often have wrinkle-free faces. Most important of all, excessive sun may predispose people to skin cancer.

The sun has been called "the greatest carcinogen" of all. Ten to 15 years of exposure to its

relentless rays can derange the normal cell structure, which then reproduces itself along abnormal lines. In the United States cancer of the skin is the most common of all cancers, totaling 300,000 new cases a year. Fortunately, except for the form called malignant melanoma, it is the least virulent. Skin cancers have relatively little tendency to invade other tissue, and they are also readily seen and promptly treated. The two most common types are called squamous-cell carcinoma and basal-cell carcinoma. Both are increasing, and both are associated with too much sunlight.

Skin cancers are found at all latitudes and in all peoples. But the rate among relatively unprotected whites doubles with each eight to 11 degrees of latitude as we approach the equator. Epidemiologically, the cases fall into a pattern. As might be expected, they are most common among outdoor workers, such as farmers, sailors, or ranchers, and occur most frequently on parts of the body constantly exposed to the sun. One survey of 840 skin cancers showed that over 90 percent developed on the face, particularly around the thin skin of the nose. Almost twice as many appeared on the exposed lower eyelid as on the protected upper lid. They were more common on the lower legs of women, below the skirt line, and on the ears of men, whose haircuts at the time of the survey did not protect their ears. Men had cancers on the neck, back, and chest, women rarely. When women at French beaches went topless, many cases of cancer of the pigmented areolae of the breasts occurred. A skin cancer on the almost never exposed buttocks was uncommon enough to appear in a medical journal.

Both kinds of skin cancers are most common among people over 50 and those with light skin who have an inborn deficiency of melanin production that keeps them from tanning easily. The Scots, Welsh, Bretons, Irish, and other Celts are predominant in this group. They take their susceptibility with them when they migrate to other countries, and pass it on to their descendants.

One study compared skin cancer rates in County Galway, Ireland, with those of Cabooture, Australia, largely settled by persons of Celtic stock, and El Paso, Texas, with its many dark-skinned Hispanic residents. Although El Paso had more sunshine and was in a more vulnerable latitude, both the Irish and Australians had higher skin-cancer rates.

As fair-skinned people have flocked to sunnier climes, the distribution of skin cancer has changed. The Australian white population has the highest skin-cancer rate in the world. And cancers are being diagnosed in anatomical parts rarely recorded in more modest days. No thanks to bikinis and more time for sunbathing, they are now seen on the thighs, midriffs, and bosoms of tan-seeking women.

Xeroderma pigmentosum is a hereditary disorder in which the skin lacks the normal capacity to repair itself after even the mildest sunlight exposure. The kindest result can be disfiguring lesions; the most severe, early breast cancer and death. Those sentenced by heredity to the condition can survive only by spending their lives indoors. A fortunately milder condition is photosensitivity. The combination of certain substances applied to the skin or taken internally plus exposure to sunlight may cause severe rash or an outbreak of hives.

Some physical effects of the sun can be well documented, but the psychological ones are more elusive. Most of us feel happier on bright days than on gloomy ones, especially if one leaden sky follows another for an extended period. We feel lonely in the dark and panic if the lights unexpectedly go out. But enough people love the dark and enjoy foggy landscapes to indicate that these feelings are not universal. We can observe what the sun does to us emotionally, but not yet why it does so.

Most Americans take the sun totally for granted. The only time they worry about it is when a bad weather forecast threatens to postpone an anticipated picnic. It is different when the sun is a brief and capricious visitor, as in Sweden. Stockholm, the Swedish capital, is at nearly 60 degrees north latitude—nearly 20 degrees closer to the North Pole than is New York. In winter, when the sun has gone below the equator, dawn does not arrive until 9 a.m. and darkness falls before four. In summer, by contrast, Swedes make tennis dates at 11 p.m. and the birds twitter them awake at four o'clock in the morning.

Because their hours of sunlight are so few, Swedes cherish every minute. Sven Pedersen, a television producer for the state-owned network, lives in Stockholm, owns a weekend home in the pine-studded archipelago—and lives for the sun. "A day without sun is a day lost," he says. "My children and I look toward the summer and those days of soft, late-evening

Attitudes toward exposure to sunlight have drastically changed.

light. They are so few that we remember every one, all winter."

When Midsummer's Day in June gives Swedes 20 consecutive hours of daylight, the whole country celebrates. The longest day of the year is spent feasting and partying. And Swedish devotion to sunlight sometimes shocks more prudish nationalities. "The first bright day of spring, everyone rushes outdoors and strips their clothes off," Sven says. A few summers ago, an Iowa town got an eyeful when a Swedish tour bus passed through one bright noon. The bus stopped in a park for lunch, and the sun-loving visitors promptly undressed to catch a few warming rays.

To get them through the long winter, the Swedes have their own holiday. The festival of St. Lucia, goddess of light, ushers in the midwinter celebration of the knowledge that the sun will return. On the morning of the feast day, the youngest female child in the household rises at daybreak and, wearing a wreath of candles on her head, awakens the other members of the family. The symbolism then gives way to a huge feast which continues through Christmas and the New Year.

Years ago, only a meager handful of people, polar bears, and penguins experienced a 24-hour stretch of daylight or darkness. The contemporary world has changed all that. Oil workers on Alaska's North Slope, miners in Spitzbergen, fishermen in Arctic waters, and scientists in the Antarctic—all live routinely in remote environments of continuous summer daylight and winter darkness. Similarly, nuclear submariners live up to three months beyond the reach of natural light; space travelers ride in a world where the sun never goes down. Even Earthlings can have a brief taste of timelessness if we travel by long-distance jet and clock off 24 straight hours of daylight.

In recent years, both Arctic and Antarctic environments have provided a ready-made laboratory for observing human behavior in conditions of perpetual daylight or darkness. Scientists are unanimous that behavior changes, but are uncertain as to why. Probably response to light and darkness is mediated by the central nervous system. There is also growing evidence of a role for the little understood pineal gland, an endocrine organ attached to the brain. It secretes a mysterious hormone called melatonin, whose only known function is related to the darkening of skin. The science of chronobiology has established that some bio-logical functions follow a daily rhythm, so that certain adrenal hormones, for example, are secreted mostly at night, whereas potassium is excreted by the body most prominently during the day. The intriguing question, of course, is whether these rhythms are innate or are responses to daylight and dark.

Anecdotal observations certainly document how people act when surrounded by continuous darkness. Dr. Jay T. Shurley of the University of Oklahoma, who has conducted several studies of Antarctic personnel, says emphatically, "When people are cut off from the sun, they become anxious and depressed. The sun is their source of energy and light, and they cannot help but miss it."

The toll of the dark is difficult to quantify because other strains are involved. An oil worker on Alaska's North Slope, for example, not only lives in continuous darkness but in an isolated community where winter temperatures hover at 40 below. "Teasing out" the impact of any given factor would be exceedingly difficult. In any case, crews work seven straight 12-hour days, then are flown to rejoin their families for seven days' rest. Such a schedule is not possible in the weathered-in Antarctic, multiplying the strain there. The enforced intimacy of small-group living increases the friction even more. Jerry Huffman of the National Science Foundation, who spent two winters in an 11-man station recalls, "Everyone went off by himself at least once a day," and some turned to drink or overeating. How much

Tromso residents greet the Sun with much celebration on Soldag—Sun Day.

199

the darkness contributed is unknown, "but it certainly didn't help."

One topic that has been researched in great depth is the effect of light—or its absence—on sleep. Both continuous darkness and continuous daylight disturb normal sleep. This is not a new finding: as early as World War I, night shift workers were reporting sleep problems in steady daylight. At the Poles, however, winter darkness seems to cause more problems than summer daylight. In two studies Dr. Shurley found severe sleep disturbances among winter crews. Careful monitoring shows sleep patterns of seven and a quarter to seven and a half hours in winter, seven and three-quarters to eight hours in summer.

Insomnia is such an occupational hazard in the Antarctic that the men have coined their own name for it. "I've got the 'Big Eye'," they say when they seem to be plagued by the widespread problem of sleeplessness. The ironic part is that they can go to bed almost whenever they choose in the chronic darkness. Most can make their scientific observations on any appropriate schedule and record their findings as they like. Customarily, they work until they are tired, often 16 hours at a stretch, then sleep until they wake up. They usually sleep in a single stretch of seven to eight hours. There is little catnapping between times.

Dr. Shurley, however, has found a distinct difference in sleeping rhythms. Normally there are two types of sleep—REM (Rapid Eye Movement) and non-REM—which alternate at 90-minute intervals through the night. REM is dreaming sleep, characterized by rigid muscles and twitching eyes; non-REM is basically dreamless. Non-REM is classified in four stages corresponding to successively deeper levels of sleep. When we fall asleep we plunge quickly through these stages to stage four, "dead to the world" sleep, after which we gradually reverse the process until we move from stage one into REM.

In the Antarctic, Dr. Shurley found, sleepers never reach this deepest stage four sleep at all. The stage simply disappears. The lack may be responsible for some of the other behavioral changes reported, like anxiety and depression, although this has not been proven. Sleepers do go into stage four during the summer, and they return to normal when they go home, apparently none the worse for wear.

Independent and totally unscientific observers suggest that the deepest level of all occurs just after first light—and just after the alarm clock stops ringing.

Throughout history, the sun has been man's timepiece, telling him when to sleep, when to wake, when to work, when to eat. Man has augmented that timekeeping function by forcing the sun into uniform 24-hour days. The sun, however, does not follow the clock. And when we ignore the clock and the sun is not present to give its cue, a curious thing happens. The daily cycle lengthens, and body time overruns clock time. In experiments in Antarctica, Norway, Spitzbergen, Greenland, and Baffin Island, subjects have established for themselves a longer-than-normal daily cycle of about 25 hours and 15 minutes. The body functions have adjusted themselves to this new internal clock.

This observation has been confirmed in laboratory settings where time is suspended. Dr. Wilse Webb of the University of Florida has tested subjects on arbitrary "days" ranging in length from three to 48 hours. The subjects sleep, eat, wake, and work without benefit of time cues, according to a schedule worked out by the experimenters. In every case, when the subject is allowed to "free run," Webb says, he falls into a 25-hour pattern.

Michael Sifre, a French physiologist, tried to follow a life without time for six months in a cave near Del Rio, Texas. He followed his own dictates while workers on the surface recorded his activities without regard to the clock. The results showed that Sifre's "days" ranged from 20 to 48 hours and that often he kept "double days," two normal days long. The scrambling of his body rhythms, Sifre wrote later, so affected him that he did not return to normal for months afterwards.

Men who endure the Antarctic night usually do so only once in a lifetime. But to 40,000 residents of Tromso, Norway, a fishing and oil-producing town 200 miles inside the Arctic Circle, alternating periods of "midnight sun" and continuous darkness are a way of life. Tromso, the northernmost permanently populated community of any size in the world, huddles on the ragged fjords and mountainous islands of Norway's Arctic Coast. The sun sets in Tromso in late November and does not reappear until late January. The residents call these sunless two months mørketiden—the murky time.

What mørketiden does to Tromso can be measured in statistics. Mental illness, physical sickness, domestic troubles, arrests for fighting,

and suicides all increase. There are more documented cases of drug abuse and alcoholism. Liquor sales increase. School records show that children's grades drop during the winter darkness. (But interestingly, fewer babies are conceived than during the continuous daylight of summer, birth records show.)

Residents report less quantifiable effects. "Mørketiden brings out the worst qualities in people," one man has said. "Envy, jealousy, suspicion. People get tense, restless, and fearful. They become preoccupied with thoughts of death and suicide. They lose the ability to concentrate and work slows down. People talk about the light constantly and long for the sun to come back."

The sun comes back to Tromso on January 29. The residents call it *Soldag*—Sun Day—and as in ancient times, it is an occasion for great rejoicing. Everyone prays for a clear day so that the thin yellow-white disk may actually be seen with the naked eye, not simply sensed through lowering winter clouds. The schools take a holiday and children parade through the streets. Traditional sun breakfasts feature sun coffee and sun chocolate, and people congratulate each other on the sun's return.

"Hey, Sam, the big round yellow thing came up again."

About 11 a.m., people begin to look toward the south; many head for the higher vantage points to catch the first glimpse of the long absent visitor. As in the Antarctic, a great thrill of anticipation ripples through the crowds as the sky first turns dimly pink and the reflected light illuminates the distant hillsides. The sky gradually brightens and the sun pokes its head between two hills, always at the same spot every year. Echoing the scientists in Antarctica, the children shout, "She's back! She's back! The sun is back!" People begin to cry. As the sun is first seen, other residents follow the tradition of opening the Bible at random to locate the verse that will guide their fortunes in the coming year. Children touch the walls where the sun strikes to bring good luck in coming months. As in Antarctica the moment ends quickly, but everyone perks up in a brighter mood.

Why do humans feel so deeply and intensely about the sun? Is it only that we are warmed by its rays, and brightened by its light? Or is there a biochemical explanation, an interaction between the rays of the sun and the body's chemical structure, altering it in a way that accounts for our moods, sleep disturbances, and depressions?

In recent years, a new scientific subspecialty has been exploring these questions. Photobiology studies the effects of light on all forms of life. It concerns itself with physical changes as well as psychological ones.

"The accepted wisdom is that light does not affect us beyond the surface because it cannot enter the body except through the eyes," says Dr. Kendric Smith of Stanford University School of Medicine, founder and first president of the American Society for Photobiology. "Yet we all know from experience that light does enter the body. If you place a lighted flashlight in your mouth, you can see its light flow through your cheeks. There is a famous painting hanging in the Louvre, *St. Joseph the Carpenter* by Georges de la Tour, in which a child is holding a lantern whose beams are passing through his hand." Even more so, sunlight suffuses the flesh. And the eyes, extension of the brain, are bathed in light.

Be it photobiological or psychological, tan or sunburn, day or night, there is little doubt that the sun still exerts the same powerful influence over the minds and bodies of man today as it did in millennia past. And if you don't believe me, just ask the folks in Tromso.

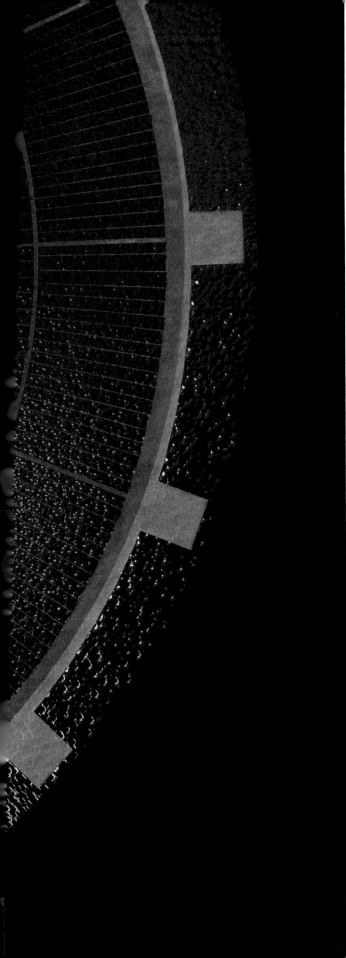

T hough just a glimmer on the horizon today, the final years of the 20th century may bask in the rays of the Golden Age of Solar Energy. But, the reader may well ask, didn't they sing something wonderful about the dawning of the Age of Aquarius, too? At least for much of the 1980s the message for consumers is *caveat emptor*, let the buyer beware. It is not that anything is wrong with the sun—we've used its energy for ages—but many mechanical solar innovations are simply not sufficiently commercialized to be good buys for the consumer on a budget. The once and perhaps future championship for perfection of design and cost effectiveness still belongs to the clothesline. There are other great bargains, though. Experts point out that an experienced home handyman can for $1000 apply conser-

Sunpower

vation and solar collection improvements that would cost $10,000 or more from a contractor. Proper insulation, a prerequisite for any efficient solar heating, is also cost effective. Observers of the technical scene frequently point out one—to coin a phrase—"blue chip" in the solar-energy futures marketplace. It is the jewel-like photovoltaic device or solar cell, one version appearing at left. We have enlarged the original to reveal its intricate beauty; its actual size is only that of a silver dollar. One of our authors, Joseph Lindmayer, is a world leader in the commercialization of solar cells—but we're getting ahead of the story.

Solar cells—or any of several other initiatives—may mature into commercial technologies. Enthralled with the inventive process, the Smithsonian Institution cheers on the dreamers and the doers. They and their forerunners have been at the task of invention almost from the moment humans could first think and act to control the environment, as recounted here by energy adviser Wilson Clark. In the beginning, as today, sunlight supplies the bulk of Earth's energy, though consolidated in such varied forms as wood, wind, coal, petroleum, elbow grease, falling water, green plants, lightning, sidewalks hot enough to fry eggs—and eggs themselves embodiments of the sun in a form we use for food.

Artist Pierre Mion speaks his mind about solar architecture. David Morris conjures up an American future where the locally based solar utilities are owned and managed within the community. Peter Glaser reveals details of his famed plan to supply the Earth with solar-generated electricity beamed down to the planet's surface from orbiting power stations.

Finally, National Air and Space Museum astronomer and guiding genius for this volume, Von Del Chamberlain, presents a sensitive Epilogue. He sums up not only the major themes of *Fire of Life*, but evokes the hopes, fears, and feelings of men and women who through the centuries have confronted the awesome sun, with all its implications for life and human destiny.

From the beginning, the availability of energy—the capacity to do work—has guided and determined the destiny of mankind, whether in small hunting bands, medieval villages, or highly complex industrialized nations.

The initial human settlements were blessed with an abundance of energy from the sun—in the direct form of radiant heat and in indirect forms such as solar energy stored in wood. With all of our sophistication and worries about the future, the revolutionary importance of the discovery of fire by *Homo erectus* can scarcely be appreciated today. Fire enabled people to establish communities—around the first genial hearths.

For sheer survival, early societies oriented their architecture around the sun and the cycles of nature. Neolithic people built houses and community structures from locally available materials—wood, earth, stone. They learned to adapt housing to the needs of local climates. Various architectural forms arose in different regions—as houses built on stilts near lakes or below ground in the desert. In some humid areas huts were woven of branches and bamboo. Hewn wood served in many climes.

One often finds rustic elegance. The great architect Le Corbusier wrote:

Look at a drawing of such a hut in a book on archaeology: here is the plan of a house, the plan of a temple. It is exactly the same attitude as you find in a Pompeian house or in a temple at Luxor . . . . There is no such thing as a primitive man; there are only primitive means.

Advances in agriculture, husbandry, and the control of water and fire gave man more sophisticated means and provided a surplus of food energy. Surplus food made it possible for people to specialize—to leave the farm to become craftsmen. Surplus energy also flowed into such complex endeavors as the construction of temples, storehouses, and fortifications.

Surplus energy, as we will see, is the key to understanding the rise—and fall—of civilizations. The rise of the city, perhaps 6000 years ago, was crowned by such technological innovations as the plow and copper metallurgy. Improvements in measurement and computation, writing, and various arts made for greater efficiency in production and yet greater surpluses. Lewis Mumford, in his comprehensive survey *The City in History*, notes that the development of the city did not eliminate earlier aspects of technology and culture, but "actually brought them together and increased their efficacy and scope."

With its productive advantages, urbanism also brought new, potentially devastating problems. Wealth, for example, invited plunder. Edward Hyams points out in *Soil and Civilization* that the tremendous soil fertility of the Mesopotamian plain, annually fertilized with alluvial silt, was ruined by the incessant wars of the Assyrians. Of the region south of Hit, he writes that it was:

. . . made by the mountains and the river, has since been unmade by man; it is a treeless, open, badly eroded soil, or rather corpse of a soil dead of *Homo militaris*, a variant of the disease organism, *Homo sapiens*.

Wilson Clark

Sun, Energy, and Civilization

Not only did new conflicts arise as more energy was harnessed, but efforts redoubled to exploit basic resources. Consider wood again. Trees played an important role in the emergence of civilization, as the great forests could be counted on not only for firewood but—in a broader sense—for the maintenance of climatic stability. Trees not only afforded protection from the fierce sun of summer, but they also provided protection against chilling winter winds. Combined with the ability of forests to add fertility to the soil and to prevent large-scale erosion, woodlands have played an immensely important stabilizing role in the development of human society.

The deforestation of the Middle East and southern Europe acts as a grim reminder of the folly in not recognizing the values of a historic, biological partnership between man and forest. In Italy, the clearing of the woods for agricul-

Frozen-in steam engine awaits summer thaw on the arctic coal mining island of Svalbard (Spitzbergen). Fossil sun energy in chemical form—coal —represents just one resource in the world's energy-supply network.

ture began in Neolithic times, about 2000 B.C. Deforestation was accelerated between 700 and 800 B.C. as wood was converted into charcoal for iron smelting. The result: a chain reaction of environmental devastation.

The same tragic cycle has been repeated many, many times, and has affected almost every land. America's Dust Bowl of the 1930s graphically illustrates the tremendous losses to society caused by massive soil erosion. In essence, what we sometimes take for gain actually results in a net energy loss and ultimate destruction of a previously undervalued

Wind rules the wine-dark seas as pirates attack a freighter, from an ancient Greek pot. In the ancient Mediterranean, sails gave men their first big mechanical prime mover—wind power being a derivative of sun energy.

source of wealth and community well-being.

The ancient Greeks, as we might expect, described the situation well and even coined a word for it. The Athenian society, according to a recent commentary by Murray Bookchin:

. . . was based on a yeomanry and on small agricultural holdings, (in which) town and country were brought into delicate balance . . . . To the Greeks, this social equilibrium was summed up by the term *autarkeia*: a concept of wholeness, material self-sufficiency, and balance that is the core of the Hellenic outlook.

This attitude certainly contributed to the development of early Greek city planning, based on the relationship of buildings, streets, temples, and the community at large to the sun, and to the demands of local climate. By the 4th century B.C., when Athens banned the production of charcoal from olive trees, solar energy was well on its way to assuming a preeminent position in determining the course of Grecian architecture and allied institutions.

In one of the earliest observations on solar architecture, recorded by the historian Xenophon, Socrates stated:

In houses with a south aspect, the sun's rays penetrate into the porticos in winter, but in summer the path of the sun is right over our heads and above the roof, so that there is shade. If, then, this is the best arrangement we should build the south side loftier, to get the winter sun, the north side lower to keep out the cold winds.

A representative dwelling, with six or more rooms on two floors, comprised a total of about 3200 square feet of floor space, by no means small by modern standards. The square Greek houses were constructed with thick adobe walls, one and a half feet thick on the north exposure, to keep out the chill of winter winds. Main living rooms, about 16 feet deep, faced a portico supported by wooden pillars on the south exposure, leading to open courts.

In winter, when the sun travels in a low arc, the Greek design permitted light to enter the home through the south-facing portico. Once inside, the radiation was trapped in the earthen floor and thick adobe. Earth and adobe provide a useful form of heat storage, and allow for a gentle release of solar warmth through the winter night. Elegant designs allowed for both the lower and upper floors to receive light.

The Romans rigorously applied Greek solar design principles, and improved upon them. In his ten-volume *De Architectura*, a still fascinating guide to applied solar concepts, the Roman architect Vitruvius laid out careful guidelines for the siting and orienting of houses, public buildings, temples, and even whole cities. He stressed that the situation and construction of the individual structure, as well as the plan of the whole community, should fit both topography and climate.

By the 1st century A.D., window glass had been developed in Rome, and this innovation made possible much more effective solar design

Top: Yugoslav grape crusher uses the ancient method to render grapes for wine. Egyptian threshing sledge, below, harks back to civilization's first vehicle, long before the wheel. Grains, the grape, honey, olives for food and oil were added to the Mediterranean larder through application of human and animal muscle—the power that helped set Western civilization in motion.

and construction. Pliny the Younger described the *heliocaminus*, "solar furnace," a room in his villa with southwest windows glazed either with mica or a crude form of glass. The idea of such a splendid heat collector was applied to agriculture as well, and the first greenhouses appeared. Romans also applied solar design principles at the public baths, places of great social importance. The hottest bath, the *caldarium*, faced the south, and enormous windows covered the entire south wall of a typical bath house. Through careful design and inspired use of building materials, solar heat was trapped in the floors and walls for storage and additional heat in the evening.

The Romans also developed the first legal concepts and laws protecting the access of buildings to sunlight, especially for the *heliocaminus*. And to this day, certain rights to unobstructed sunshine are recognized by British law and the legal codes of other lands.

Wind is another form of solar energy: air converted by nature into useful motion. Harnessing this prime mover resulted in an early energy advantage for certain nations. The story of ships and sails emerges in antiquity. Even in ancient Egypt, however, some sailing craft displaced as much as 150 tons and their sails could capture the energy equivalent of an 80-horsepower marine engine. By the time of Rome's suppression of Greece, merchant ships of 250 tons and larger were built, achieving about 120 horsepower—a far greater concentration of energy than any on

land. Exploration, trade, and colonization by sea were characteristic of many of history's most exciting and vigorous periods. By the early 1900s the greatest windjammer of them all, *Preussen*, developed the equivalent of 5000 horsepower.

How ironic that in the end the sophistication of sail in Roman hands was turned to stripping the empire of its forests to fuel the depleted environs of Rome. The "woodships," *naviculari lignarii*, made extensive runs to North Africa and France to fetch firewood for home hearths.

But let there be no mistake. In those days the primary energy convertor was the slave, and the conversion of solar energy into wood, food, or moving wind or water was accompanied by slave-power. At one time, the great market-place at Delos saw the sale of 10,000 people in a day, and at one time Rome held at least 400,000 slaves. Yet—in strictly thermodynamic terms—it was little wonder that slavery worked so well. The energy efficiency of human labor is striking even when compared to the values for domestic animals. James Watt calculated the energy produced by a horse in his day at about two-thirds horsepower, equal to six horsepower hours in a nine-hour day. A man working nine

Remnants still linger from the medieval world that brought with it an effective though limited solar society. From left to right: Egyptian felucca *sails bear witness to the spread of sea trade. Windmill sails rise on a Rembrandt landscape. French illuminated manuscript, complete with solar chariot at top, reflects what for the nobles was the happy round of feudal life. Iranian arches present an example of an architectural adaptation to a harsh desert climate.*

hours a day produces only one-quarter as much energy. The horse, however, consumes ten times more food energy than a man. So, in reality, the total delivered energy is two-and-one-half-times greater for a man.

Since slaves were so spendidly fuel efficient, why then are horses and draft animals so significant in human history? The answer lies in the fact that the horse can deliver energy in a short period of time at a far greater rate than a man working alone. And, as we will see, the concentrated release of energy has special advantages. In fact, we have already seen it in sail-power.

Major breakthroughs in energy history followed the development and spread of water mills and later of windmills. Not only did these devices significantly reduce the energy demands on both humans and draft animals, but they eventually helped launch the great Industrial Revolution.

On land the earliest powerful machines were water mills. They emerged to tap the energy of fast-moving streams in hilly areas of the Near East. The design—known as the Norse or Greek mill—differed from the typical modern variety in that the axle was vertical. At the

Woodcut from 1556 reveals a crucial energy transition. After unrestricted charcoal burning denuded forests, workers mined coal to power the Industrial Revolution.

lower end of the axle a series of scoops turned in the stream, providing mechanical power to grind corn. Pliny described such mills, noting their use in northern Italy. They were small and slow, as the millstone could turn only as fast as the wheel allowed.

Vitruvius described an alternative, a water mill with a horizontal axle and vertical wheel, which may have been based on the Persian *sagiya*, a water-lifting device consisting of pots arranged around a wheel turned by people or draft animals. Such devices were in use in Egypt in the 2nd century B.C.; the Roman modification consisted of running such a "scoop wheel" in reverse. The wheel was connected to the millstone through wooden gears. A Roman mill located at Vena, with a wheel seven feet in diameter, produced about three horsepower, and could grind about 400 pounds of corn per hour. A grain mill worked by men or donkeys could—at best—grind only 10 pounds an hour.

The real spread of windmills came several centuries after the fall of Rome and opened a new chapter in the availability of energy for agriculture and small industry. One reason for the delay was Rome's heavy dependence on slaves. For example, Emperor Vespasian (A.D. 69–79) was opposed to mills because he was afraid they would create unemployment.

By the 11th century, almost every village with access to a running stream had a water-wheel for grinding flour. The mills were of various designs and accomplished a wide range of work from tanning leather to making cloth. A tidal mill is recorded in the *Domesday Book*. It was working in Dover harbor before 1086. In all, there were at that time 5624 water mills in England south of the Trent, mostly of the design described by Vitruvius.

The earliest known windmill turned in Persia during the 7th century, perhaps a successor to the wind-powered prayer wheels of the Buddhists. Like the Greek waterwheel, the Persian wind machine had a vertical axle and typically consisted of a two-story building housing a millstone on the upper floor, with a wheel driven by six to 12 wind sails.

Western designers tilted the windmill axis from the vertical toward the horizontal, with classic Netherlands windbeaters angled slightly upward to give the sails a more efficient chop at the breeze. A Norman deed in 1180 describes the earliest type of horizontal axis windmill, and such devices were common in North

Europe by the end of the 13th century. By the 18th century, the English had developed sophisticated mills with small fantails, wind-powered devices to help swivel the big wooden windmill blades into the wind.

By harvesting sunshine directly for crops and house heating, and indirectly with water and wind, Europe created a truly solar-based industrial civilization between the 11th and the 18th centuries. Each new technology was carefully tailored to make the most of the newly discovered energy sources. Water mills and windmills averaged an energy contribution to society of only five to ten horsepower per installation, yet even this modest output was enough to influence profoundly Europe's economic and industrial patterns.

The dispersed power of the sun's energy, derived from agriculture, water, and wind, was soon to give way to a more concentrated source of energy—fossil fuel.

Between 1698, when Thomas Savery patented the first prime mover powered by steam, and 1782, the year James Watt obtained a patent on a vastly improved steam engine, civilization entered the era of industrialization, centralization, and dependence on fossil fuels—solar

energy stored in the earth in the form of coal and oil. The inventors of the early engines knew little about theoretical refinements and efficiencies. They were interested in applying brute force to pump water from mines so they could get more coal in use quickly.

Understanding the relationships between the new engines and their fuel sources was important, though, if better devices were to emerge. And, indeed, the principles of thermodynamics were being formulated by the early 19th century. The science of thermodynamics—from the Greek roots *therme* or "heat" and *dynamis* or "power"—defines the relationship of energy, heat, and work.

In 1819, a young French artillery officer and physicist, Sadi Carnot, published a key technical paper which elucidated the principles governing the *cycle* of energetics in steam engines. He found that the best engines accept steam at the highest possible temperature and reject it at the lowest, and he conjectured that a "perfect" engine would have to be capable of operating with no energy losses. Later, the German physicist Rudolph Clausius refined Carnot's ideas and reasoned that a heat engine would convert some of the original energy content of the fuel

Below, gala opening of the Brooklyn Bridge in 1883 marks a poignant moment in the spread of modern, energy-intensive society across the globe. Already, sail is being replaced by steam for water transport. The steel-suspended span indicates burgeoning industrial prowess. Right: topsy-turvy specter of man and oil derrick suggests what to many observers seems like an almost heroin-like technical addiction of industrial nations to the quick energy fix of oil. At far right, waste gas flares from an Arabian oil production rig.

to useful work in the engine, then dissipate waste heat. In keeping with English physicist James Joule's law that energy and matter can neither be created nor destroyed, no energy was destroyed in the process, but that which remained lost its usefulness—a condition for which he coined the word entropy. He went on to describe the wider significance of his ideas about this disturbing transformation: "the energy of the universe is constant, but the entropy of the universe increases toward a maximum."

The extended significance of this law is noted by economist Nicholas Georgescu-Roegen:

The Entropy Law teaches us that the rule of biological life and, in man's case, of its economic continuation is far harsher. In entropy terms, the cost of any biological or economic enterprise is always greater than the product.

The laws of thermodynamics were identified just as the Industrial Revolution was getting up steam and the various engines of commerce and industry were clanking and hissing their way across the European continent and beyond. As the laws were better understood, bigger and more powerful engines could be built, and, with their "energy conservation" increased, more and more useful work could be wrested from them per pound of fuel consumed.

This key lesson of conservation and efficiency can be illustrated well by a British venture of the mid-1800s. The *Great Eastern*, a 19,000-ton vessel, was to have proved England's industrial prowess. The ship, equipped with bunkers holding 12,000 tons of coal, was to voyage to Australia and back without refueling. Practice couldn't match theory, however. The ship would require 75 percent more coal than her coal storage capacity—more coal, in fact,

211

than the weight of the ship herself.

To my mind the *Great Eastern*'s story can be taken as a parable, a symbol of the early Industrial Revolution, a period of expansion characterized by the liberation and use of massive quantities of energy and other physical resources. Earlier civilizations, in constant peril of food and energy shortages caused by natural events, were rarely able to accumulate surplus energy. The new industrial age gloried in surplus energy, wallowed in energy supplied by the Earth's fossil dowry.

If coal fired the furnaces of 19th-century European industry and commerce, in younger, less exploited America seemingly limitless forests provided fuel for increasing industrial expansion. And although wood was indeed abundant, Americans were nothing less than profligate in its use.

The American attitude toward wood was characterized by R.G. Lillard, in his account of the pioneer:

The kind of hospitable settler who burned a whole log in order to boil a kettle of tea didn't consider his fire psychologically good until he had crammed a quarter of a cord into a space eight feet wide and four feet deep and had a small-scale forest fire roaring in front of him.

As vast tracts of forest fueled the United States, extensive coal deposits were located and developed. Soon oil, then gas and electricity replaced whale oil for lighting, while coal became the primary industrial energy source. Hydropower was another significant source of energy. The mills of New England, birthplace of America's industrial boom, were largely powered by water, and the Dupont Company's powder mills were driven by the Brandywine River for most of the 19th century.

Electricity produced from falling water made its debut at Niagara Falls in 1882, just 26 days after Thomas Edison opened the first U.S. electric power plant (coal-fired) in New York City. Surprisingly, hydroelectric power represented less than three percent of the overall energy sources in use in the U.S. in 1900. The dominant energy sources were wood and coal; even at the close of World War I, coal's share of the national energy market was six times that of petroleum products. Another key solar source had dwindled by the 20th century—the use of wind power. In 1850 the use of windmills represented about 1.4 billion horsepower of work.

The equivalent amount of coal required to produce this much energy is 11,830,000 tons. Yet by the 1870s only half this much wind power—mostly used for pumping water on the prairies—was in use in America. The modern fossil era had arrived.

Since World War I, the availability of massive quantities of surplus energy from fossil fuels, and, more recently, from nuclear fission, has made possible a quantum leap in mechanization affecting almost all aspects of human life in the Western world. Agricultural practice in the U.S. provides an example. To harvest an acre of wheat in the 1880s took 20 man-hours, yet by the 1930s, only six hours were required.

Yes, we bought time, but we paid, and will pay dearly. Perhaps the central problem for modern society is its heroin-like dependence on increasingly scarce, nonrenewable sources of energy. A leading petroleum geologist, M. King Hubbert, points out that the time required to deplete a full 80 percent of America's oil and gas resources "is the 61 year period from 1939 to the year 2000, well under a human lifespan."

Most of the cheap and accessible energy deposits have already been tapped; the fuel resources of the future will require more and

Mesa Verde cliff dwellings received sun in winter, shade in summer. Placement of structures remains an important principle of passive solar heating. "Sun-Maid" raisin girl represents an age-old process of preserving produce by drying. The playful "Do-Nothing Machine," built in 1957 by Charles Eames, shows early application of solar cells.

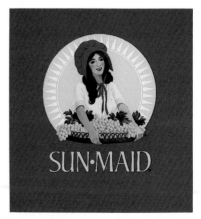

more energy for the process of exploitation. As this energy is used, there will be less and less available, less *net energy*, for society. Thus, the entropy factor, elucidated in the early years of the 19th century, haunts our future.

What then, of solar energy? I suggest that as energy costs and consumer prices move ever higher, many of the older, localized patterns of civilization will re-emerge. Many energy sources, including high-technology applications of solar energy, will be available in the future, but costs will be high and energy will not be so widely available as in the extravagant era of the 1970s and early 1980s. I leave it to the articles that follow to explore the solar future in depth. I will only mention here that a variety of housing designs exists, including active solar systems that utilize heat storage and summer cooling techniques. Whole communities have recently sprung up that make the most of effec-tive geographical siting and precise solar orien-tation of individual structures.

A fossil-fueled city can be thrown up any-where, but context must be considered if you want to go solar. The advice of Sir Ebenezer Howard is still timely. In 1902 he called for the development of "garden cities" which would relieve urban congestion by dispersing the human population and make use of localized industries and agriculture in the communities.

Hitherto in the American experience, only the southwestern Indian pueblo communities of 900 A.D. have shown the level of sophistica-tion in solar design just now reappearing in our civilization. Let us hope that the lessons taught by past cultures will be heeded and, as Ebenezer Howard dreamed, "Town and country must be married, and out of this joyous union will spring a new hope, a new life, a new civiliza-tion."

Designed specifically to save fuel, a solar community emerges—conceived and painted by the author. Narrow streets retain less summer heat, so homes require less cooling, while larger green areas can be devoted to shared backyard or commons areas. This encourages outdoor living, and plantings provide a buffer against hot and cold winds. Produce can be easily carried or biked out of shared garden space, far left, thus saving fuel; otherwise much food would be transported long distances.

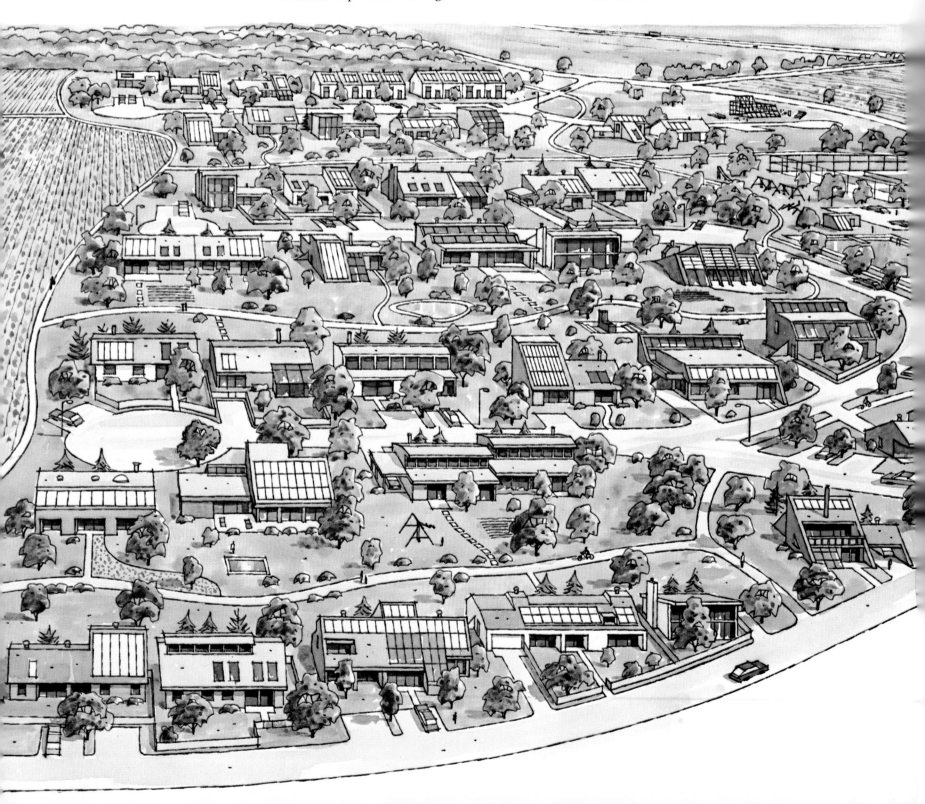

Pierre Mion

Solar Communities

Editors' Note: *As a result of extensive research for technical and scientific artwork, Pierre Mion has gained wide experience on which to base his comments about solar hearth and home. Here, Mr. Mion speaks his mind about energy-efficient living and also illustrates his own story.*

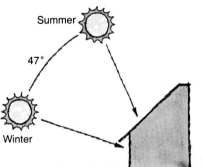

Overhang shades windows and walls against high summer sun. But in winter, when the noonday sun lies more than 40° below summer zenith, heat-filled rays enter the structure. This simple architectural adaptation—fully exploited by the Greeks and Romans—is still an energy saver.

Summer

47°

Winter

All around Washington, D.C., I see construction under way on tract after tract of the same old conventional housing. New homes face every which way. Carpenters use two-by-four-inch studding instead of two-by-six, which provides more insulation. Dark roofs absorb summer heat. Single-pane glass allows escape of valuable heat in winter, and north and west walls are full of windows. Conventional heating systems are used to do all the work of warming and cooling.

These are but a few examples of how too many builders everywhere ignore proven principles that have been with us for thousands of years, and until fossil fuels were discovered were used far more than today. One of the most important consists of little more than orienting our buildings to take advantage of the sun in winter and shade in summer.

More simply put, most modern houses are gas guzzlers. For some reason, many architects and contractors, especially on the Atlantic seaboard, act as if they believe that there is no energy crisis. Worse yet, if they do believe there is one, they don't seem to care. It's just simpler and cheaper in the short run to continue building the standard American box. And commercial buildings are still being constructed with the same lack of concern.

In my opinion there should be a national building code prohibiting any new construction that is not fuel efficient and in accord with solar principles. When fossil-fuel energy was less expensive and, as we blindly thought, inexhaustible, it made little difference how we built our homes. Now, presumably, we should know better. But, except for a few far-sighted planners, we still follow the traditions of a wasteful era that we'd do well to put behind us. Now is the time to rediscover the sun.

Contrary to popular belief, a solar house need not be complex or expensive to build. A few simple guidelines apply. In the Northern Hemisphere the building must face the south for maximum exposure to the sun, and this should be the house's longest side. In the winter, with the sun shining from a low southerly angle, large glass areas will admit hours of solar radiation. As sunlight passes through glass it strikes various surfaces. Part of the visible light will reflect back out of the glass while the heat is trapped inside. This principle has been called the greenhouse effect.

As you may already experience in your own home, a south-facing room is always the warmest during the winter months. And if your house has a good expanse of southern exposure, you may already be halfway to solar heating. All that is needed to make the room more thermally effective is an efficient system to store the sun's energy.

In summer, when the sun is at a much higher angle, roof overhangs, recessed windows, or movable awnings and shutters block most solar radiation from reaching the interior of the house. A light-colored roof reflects much of the sun's heat away from the building. Such simple design features contribute to what is known as a *passive* solar system, in which no mechanical devices are necessary to collect, store, and distribute the sun's energy.

An *active* solar system employs large roof collectors that permit the sun to heat either air or a working fluid which is then pumped to a storage unit. These heat reservoirs are usually in the basement or crawl space because of their large size and weight.

Two types of storage units are commonly in use. The hot air type consists of a large enclosed bin filled with rocks. Air warmed by the sun is pumped from collectors on the roof or in the yard to the bin, where the rocks absorb the heat. Warm air can then be pumped from the bin to the various rooms through conventional ducting. In the other system, water is heated in solar collectors, then stored in a well-insulated tank. The warm water releases its heat through standard radiators.

High vents exhaust summer heat

Wing-wall to protect from sun and wind

Deciduous tree for summer cooling

House of passive solar design— so called because it requires no motors, fans, or pumps—is partially buried and constructed of thick adobe—with the internal Trombe wall design, see dia- *grams at right. Top tier of windows heats the separate rear of the house, while Trombe walls behind windows provide heat for the front.*

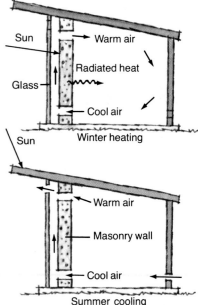

Trombe-Wall Construction

Sun

Glass

Warm air

Radiated heat

Cool air

Winter heating

Sun

Warm air

Masonry wall

Cool air

Summer cooling

216

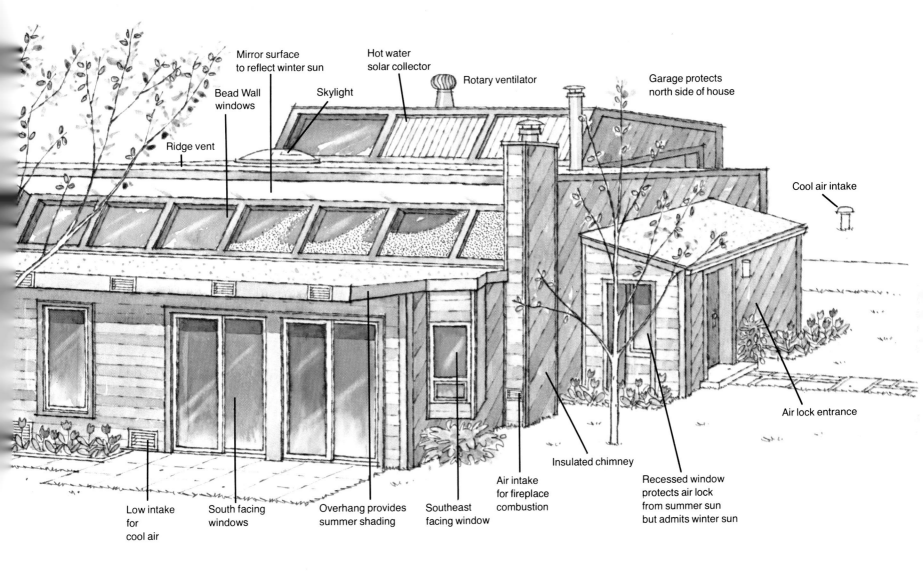

Mirror surface
to reflect winter sun

Hot water
solar collector

Rotary ventilator

Garage protects
north side of house

Bead Wall
windows

Skylight

Cool air intake

Ridge vent

Air lock entrance

Low intake
for
cool air

South facing
windows

Overhang provides
summer shading

Southeast
facing window

Air intake
for fireplace
combustion

Insulated chimney

Recessed window
protects air lock
from summer sun
but admits winter sun

Hybrid solar house, above, combines a few active solar systems with a basic passive design, as opposite. Upper windows have two panes separated by four inches, and at night polystyrene beads fill the space and keep heat inside. The procedure is reversed in summer. A southeast facing window is angled to catch early winter light and thus start rewarming for the day. As with the Trombe wall, left, thick walls in the cutaway at right retain heat, also help keep house cool. House at right employs a back-up furnace.

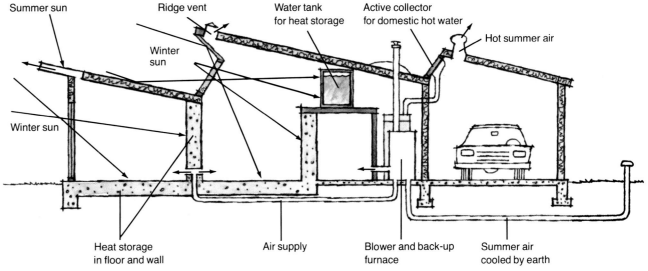

Summer sun

Ridge vent

Water tank
for heat storage

Active collector
for domestic hot water

Hot summer air

Winter
sun

Winter sun

Heat storage
in floor and wall

Air supply

Blower and back-up
furnace

Summer air
cooled by earth

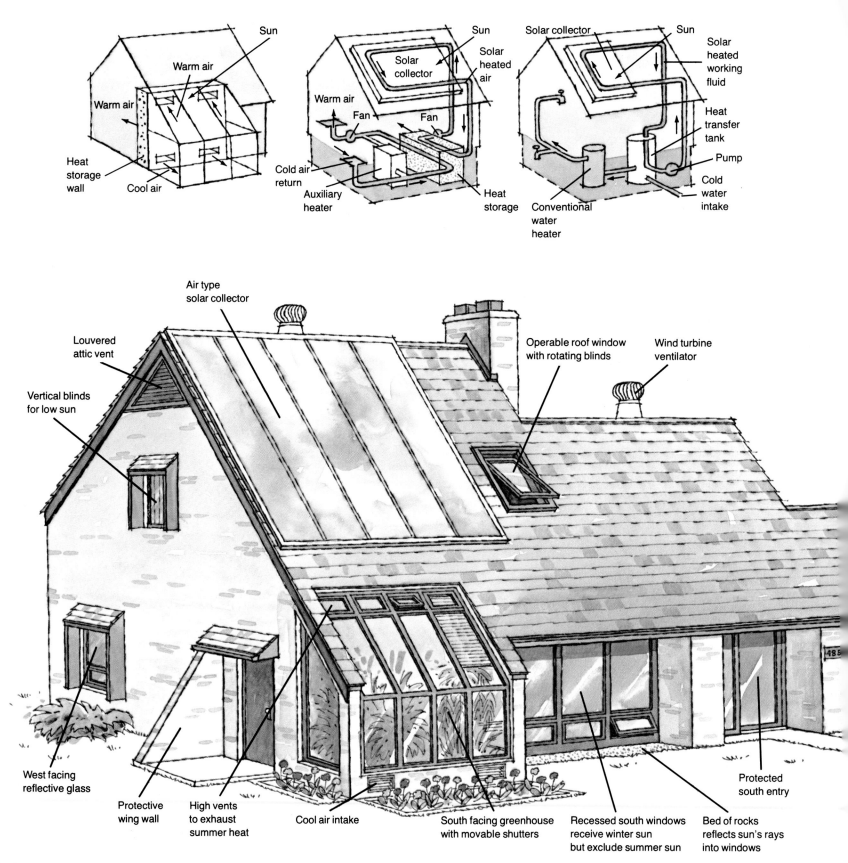

Sun

Warm air

Warm air

Heat storage wall

Cool air

Sun

Solar collector

Solar heated air

Warm air

Fan

Fan

Cold air return

Auxiliary heater

Heat storage

Solar collector

Sun

Solar heated working fluid

Heat transfer tank

Pump

Cold water intake

Conventional water heater

Air type solar collector

Louvered attic vent

Vertical blinds for low sun

Operable roof window with rotating blinds

Wind turbine ventilator

West facing reflective glass

Protective wing wall

High vents to exhaust summer heat

Cool air intake

South facing greenhouse with movable shutters

Recessed south windows receive winter sun but exclude summer sun

Bed of rocks reflects sun's rays into windows

Protected south entry

Below, left: active solar house features a blower-operated device called an air type collector. Large south-facing windows and a greenhouse augment the active solar system. From far left, diagrams show an integral greenhouse, air type collector, and a solar heater for domestic hot water. Heating by any solar system tends to be gentle and continuous, in contrast to the intermittent blasts of hot air from a conventional forced-air furnace. Whatever the system used, for comfortable living conservation features are required. And while these are best incorporated in original construction, many items—such as insulation and improved caulking—can be retrofitted as detailed on the following pages.

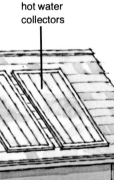

Liquid type
hot water
collectors

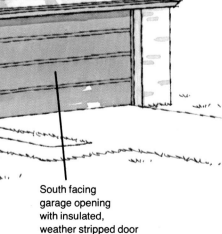

South facing
garage opening
with insulated,
weather stripped door

Most solar houses, except in some of the warmer areas of the Sun Belt, such as Southern California or Florida, utilize a hybrid combination of active and passive systems. Active houses still must face toward the south; so in addition to roof collectors, large glass areas are included. These windows or greenhouse walls are passive collectors. Most passive houses incorporate an active liquid collector for the family's hot water needs, although there are some passive designs for this function as well.

Almost all areas of the country will require a certain percentage of back-up heating to maintain a comfortable temperature during long periods of cloudy or extremely cold weather. Back-up heating systems can be the conventionally active furnaces we have in our homes today. Heat pumps are sometimes more efficient than other kinds of furnaces. But surprisingly, although winter temperatures may be low at times, most sections of the United States are quite sunny all winter. A passive solar home in Boston, Massachusetts, for instance, can be heated solely by the sun 56 percent of the time. In the Southwest, nearly all heating can be acquired by a passive solar system.

Fortunately, a variety of designs for solar houses exists, each fitted to particular geographic and aesthetic requirements. Prevailing climate is an important consideration. In cloudier or colder areas active collectors may offer the best heating system, as they can transport the sun's heat to large, remote storage areas. These can conserve the heat for longer

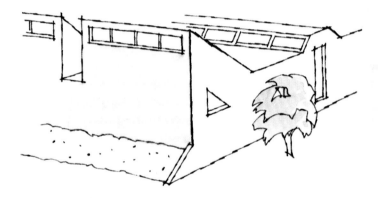

Above, in front and rear view: office building gains heat from sunlight falling on trough-like parabolic collectors. High windows admit or exclude light by freon-activated louvers that open and close automatically. Rear Bead Wall windows admit daylight for studios, then fill with beads at night to retain building heat. Window space on the cold west side is limited to tall stairwell panes, and front windows are recessed for summer shading. Earth berm in rear provides additional thermal protection.

periods than the thick wall or floor in a single-dwelling passive system. The Swedes and Canadians are now gathering solar heat during summer and storing it underground for winter. Nonetheless, passive systems require little or no maintenance and are better used wherever possible. And, all of these design principles apply equally as well to commercial buildings.

Let's be practical. Solar buildings do cost more to construct. In addition, active solar designs require periodic maintenance and eventual replacement of some equipment. Yet the long-range savings in fuel bills can be substantial, particularly if you can obtain an effective system at a reasonable price.

Going solar won't bring you a bonanza. Remaining dependent on fossil fuel won't send you to the poorhouse right away, either. But once the costs of converting to solar heat are paid, your bills will go down instead of up, and will stay down. It's a bit like possessing a small oil well right in your own backyard.

Checklist of likely ways to increase energy efficiency of a city or suburban home. You can make some improvements while others require a contractor. In all cases check local housing codes.

1. Insulate basement walls
2. Insulate water heater
3. Install electric ignition for furnace
4. Install furnace flue damper
5. Recover clothes-dryer heat
6. Insulate ducts
7. Use 3-way light bulbs, or fluorescent
8. Use glass fireplace screen
9. Use timer-thermostat
10. Caulk around windows, doors
11. Add air-lock vestibule
12. Install dimmer switches
13. Recover heat from flues
14. Double or triple glaze
15. Add storm windows
16. Install awnings for summer
17. Add solar heater
18. Install rotary ventilator
19. Install light-colored shingles
20. Install ridge ventilator
21. Seal, insulate vent pipes
22. Use exhaust fan with adjustable vent
23. Install insulation to 12-inch minimum
24. Use fluorescent lights
25. Install water-saving toilet
26. Erect wing wall
27. Install double glass with adjustable blinds between panes
28. Recover heat from refrigerator compressor
29. Install reflective windows on west side
30. Add vertical blinds to west side
31. Insulate floor over crawl space
32. Vent crawl spaces
33. Exhaust heat from refrigerator
34. Shade air conditioner
35. Face garage door into sun to help melt snow
36. Add fence to deflect north wind
37. Enclose carport
38. Plant evergreens for wind protection
39. Erect wind-driven generator
40. Plant deciduous trees for summer shade
41. Attic exhaust fan with adjustable louvers

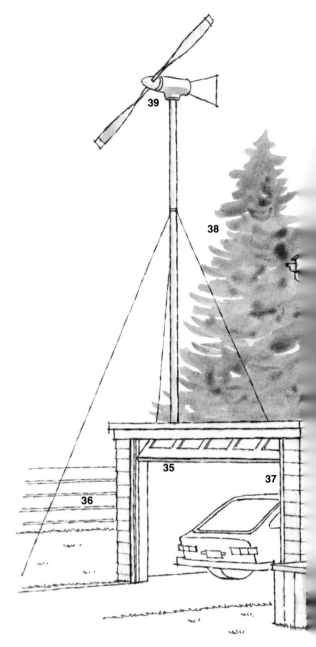

Areas of Heat Loss

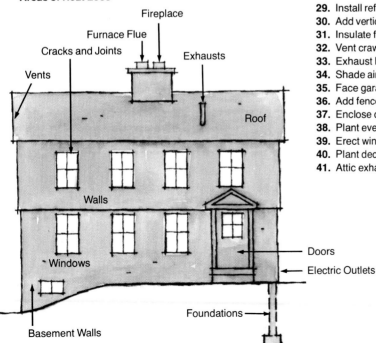

Fireplace
Furnace Flue
Cracks and Joints
Exhausts
Vents
Roof
Walls
Windows
Doors
Electric Outlets
Foundations
Basement Walls

220

Joseph Lindmayer

An Inventor's View

A criticism frequently leveled against solar energy conversion devices such as waterwheels, solar heaters, and even the power tower shown here is that they produce a low grade of energy. Though technically valid, such a criticism is, in my view, beside the point. What is significant is this: so much energy surrounds us that any common convertor can trap some and make it useful.

We really should have a greater respect for nature, the source of all our energy. As I see it, the solutions to our energy problems lie in discovering nature's secrets and using them both cleverly and wisely. For instance, phenomena exist which can convert light directly into high-grade energy; the most important being the photovoltaic effect which allows us to transform light into electricity. Significantly, this method of conversion requires no moving parts whatsoever. So, while several of the pictures that accompany this article illustrate other types of solar collectors, I will confine my remarks to the basic photovoltaic device, the so-called solar cell.

As a physicist, I worked my way through the semiconductor revolution of the 1960s and, since the technology is similar, by 1970 I found

Sunlight becomes sunpower when rays from dozens of mirrors converge at a high port in the side of a concrete tower near Albuquerque, New Mexico. The experimental project melts steel plates, tests materials, and may lead to a commercial technology for generating electricity from sunshine.

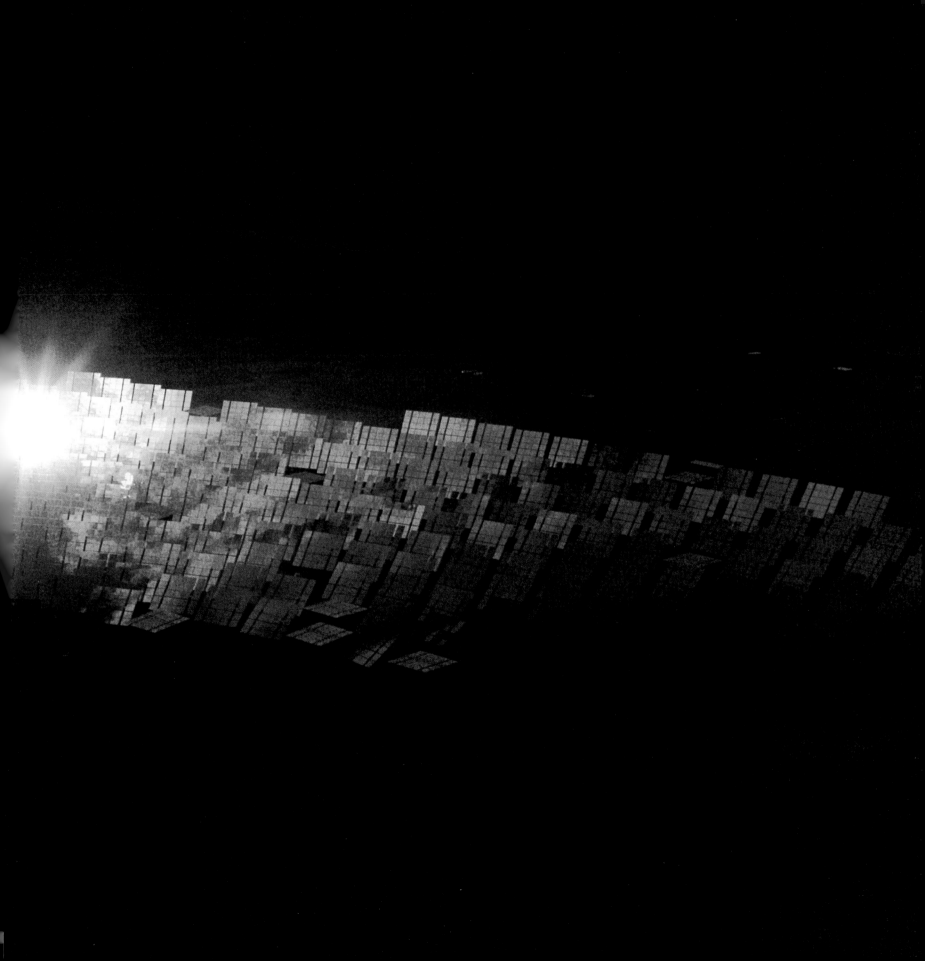

myself looking closely at solar cells. Satellites in orbit were already providing intercontinental communications, and most of them used solar cells to produce their electricity.

From my vantage point it appeared that the cells of the 1960s were overly expensive and somewhat primitive. Because of their very narrow application in space work, the early devices missed out on other developments and therefore never reached their full potential.

With the help of the broader view my background provided, I could soon announce a development called the "violet cell." It improved conversion efficiency some 50 percent and found wide acceptance in the space industry.

But the question arose: can these

solar cells be used on Earth for energy generation—and in an economic manner? The answer was immediately obvious. At 1972 cost levels, photovoltaic power on Earth would be completely impractical. Solar cells were generating electric power at a thousand times the price of household current delivered by the electric company. Yet in the early '70s, a few people had begun to recognize that the cost of energy was rising. Several voices urged the necessity of developing alternate energy sources—including renewable solar ones.

Just before the 1973 oil embargo, some others and I concurred that commercialization of photovoltaics had become feasible and should be actively pursued. I was convinced that

the proper technology for inexpensive photovoltaics could somehow be developed. These convictions prompted the formation of the first independent enterprise in the solar electric field. And as I see it, the energy realities that applied in 1973 are still with us, and they need to be more widely known and appreciated by the general public.

The dimension of the energy problem is better understood if we define some units for measuring energy. A suitable energy unit, and a very large one, is called the Quad. One Quad is equivalent to one thousand trillion British Thermal Units or 10^{15} BTUs. The United States, which uses more energy than any other nation, consumes 70–80 Quads every year. Therefore, if the whole world is to have a high living standard, thousands of Quads of energy will be required year after year.

Conventional sources—fossil fuels— can supply such large amounts of energy for only a few more decades before they run out, and therein lies the essence of our so-called energy crisis. But if we can extend our thinking and our perspective, it is very clear that the energy crisis is not real. There is plenty of energy on Earth if we focus our attention on the renewable energy sources and particularly on the sun. We cannot forget that the sun delivers about a million Quads to Earth every year and about 44,000 Quads to the continental United States alone. Taking this last figure and assuming that conversion of sunlight to useful energy could be done at 20 percent efficiency, we have about 8000 Quads.

Clearly, as little as one percent of the land could supply all of the energy needs of the United States. But even this statement is somewhat misleading, because areas already in use for other purposes can be reutilized. For example, more energy in the form of sunlight probably falls on the roof of your house than you purchase each year through the gas, oil, or electric companies. Cover your roof with solar cells and you can be your own utility.

Many of us have seen the little blue solar cells. But what is this invisible photovoltaic effect that in such a magic way converts light into electricity? First, it may be said that the photovoltaic effect is a natural phenomenon which was detected more than a hundred years ago. Scientists working in the early 20th century, including Einstein himself, investigated the nature of light and the interaction of light with matter, and by the 1930s advances in semiconductor theory lead to the understanding of the photovoltaic effect and the solar cell.

I should point out that in order to observe the photovoltaic effect, two basic conditions must be met. The first condition is almost trivial, namely that light must be absorbed by the

Mirrors that beamed light at the power tower on pages 222 and 223 now reflect the structure, opposite. Polished metal plates, below, collect a big dishful of New Mexico sky and clouds. Both solar devices gather scattered sunshine and concentrate it. A water-cooled case, opposite, holds a cookie-sized photovoltaic cell. Electricity flows when a focus of sunlight hits the device. The cell, identical to the one on pages 202 and 203, was specially designed to accept concentrated light.

225

material involved. Absorption means that the incoming photons, which are like particles of light, collide with electrons and transfer their energy to them. The electron that has absorbed the photon is an excited electron. As long as it remains so, a pair of electrical charges—one negative, one positive—exists in the electron's neighborhood.

The negative charge that we observe is the excited electron itself, while the positive charge is the place, or "hole," where the electron used to be. It so happens that the electron remains in the excited state for a short time—just how long depending on the material that holds, or hosts, the electron. In semiconductors (a class of materials between insulators and conductors)

the electron's high-energy lifetime lasts about one millionth of a second. These electron-hole pairs last only a trillionth of a second in metals. If nothing is done to the "hot" electrons, they will return from this excited state to their ground state, or, as one can say, they will recombine. In the process of recombination, the electrons lose their extra energy, transferring it to the atomic structure of the host material. This warms up the material and the whole process is the basic reason why most materials heat up under the sun. So sunpower actually exists as electricity in a material before it is transformed into heat.

More is going on here than we can sense, so let us go back to take a closer look at that photon of light as it strikes

Opposite: enlargement reveals the elegant structure common to a series of large solar cells developed by the author. The silvery trident is part of a grid that conducts solar electricity, and is applied by dipping a prepared base of crystalline silicon into a bath of molten solder. Fabrication of cells was formerly tedious and expensive because of materials and hand labor. With the process simplified and

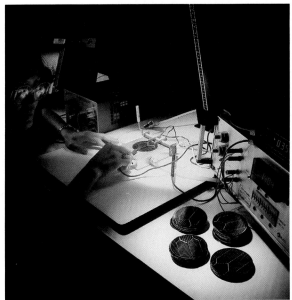

largely automated, the cost of solar electricity is expected to drop sharply through the mass production of cells. The final delicate step, quality control testing, is still best done by human hand and eye, above.

material and is absorbed. First, as we have seen, an excited electron will appear for about a millionth of a second. Its missing place, the "hole," is the positive component in the charge pair. If we do nothing to this electron within this one millionth of a second, the energy turns into heat, as pointed out earlier. On the other hand, if we could provide an internal force to separate the positive and negative charges, we could collect them with metal electrodes and generate electricity.

How could we readily provide such a force? Well, nature has been kind in this respect, and has provided a built-in electrical field where different materials touch—scientists call this a junction. Furthermore, we have discovered that the electrical field is particularly large in certain material combinations. Where cadmium sulfide contacts copper sulfide, for instance, a rather efficient solar cell is formed. Another semiconductor material, gallium arsenide, can transform nearly 20 percent of the light energy falling upon it into

electric power. But silicon is the most important photovoltaic material of all. Extremely abundant, it makes up a fourth of the Earth's crust. Also, this chemical element is harmless. Quartz and sand are basically silicon chemically combined with oxygen. To obtain the hard, gray chemical element needed for solar cells one must first remove the oxygen. Purified and melted, silicon itself can be encouraged to form a large crystal—referred to commercially as an ingot. This expensive material made from cheap quartz sand is sliced up into wafers for making transistors and integrated circuits—or chips—and also solar cells.

Slices of pure silicon produce the built-in electric field when impurities of certain kinds are added or "doped." For example, if the original ingot is doped with small amounts of boron, the silicon becomes rich in positive "holes." Accordingly, the wafer becomes a p-type. When the wafer is subsequently heated in a gaseous atmosphere containing phosphorus,

this impurity will diffuse into the wafer. The phosphorus-doped region will be very rich in electrons (which are always negatively charged) so it can be called an n-type.

The junction then is the region where the phosphorus-doped silicon meets the boron-doped silicon, forming the p-n junction that provides the required built-in electric field. This kind of technology is widely practiced, with the resulting solar cells achieving efficiencies as high as 18 percent.

Earlier types of solar cells were often disk shaped and seldom as large as a silver dollar. But newer ones cast in rectangular crucibles can be much larger. In fact, some measure five or ten inches on a side. Each cell can be viewed as a "solar battery" that produces electric power as soon as it is illuminated. A single solar cell generates approximately half a volt. The amperage varies, however, depending on the size of the cell. To produce large voltages, we interconnect cells, and the resulting solar panels provide

Architectural proposal, left, reveals Joseph Lindmayer's concept for a Solar Breeder Reactor planned for the 1980s. It will be the first factory to operate entirely from power supplied by solar cells, and its product will be more solar cells. Lindmayer estimates that during their first year of operation, cells on the big south facing wall will pay for themselves, and from then on electricity for the factory will be free. Lindmayer also thinks that the cells will probably outlast the building. Top right: Egyptian scientist erects an experimental solar array on the roof of a village house in the Nile Delta. The solar cells run a television set and bring the settlement its first electricity. In suburban Chicago, lower right, a large array powers pumps and other equipment at a gasoline station. The small building in the background supports 5184 semicrystalline cells that supply a peak output of five kilowatts.

direct current power at the same level for many years.

In some applications large solar panels are directly connected to a motor, light, television set, or other electrical load. In water pumping, for example, these panels drive electric motors for as long as the sun shines. In other cases, the solar panels are connected to battery banks that build up a power supply, thereby making available an even distribution of electricity day and night throughout the year. In this fashion, independent electric power is quite possible. Ultimately, photovoltaic power will be converted into hydrogen or even into liquid fuels.

Considering all the advantages of photovoltaic power, one may ask why it is not available on a much wider scale. One difficulty is the unfamiliarity of the potential user with this new source of energy. At the same time, the cost of this energy remains relatively high. But the price is dropping rapidly as gains in industrialization occur. We must remember, though, that when photovoltaic power was used primarily in space satellites, the industrial capacity was limited to an annual production of cells capable of providing a total of some 50 kilowatts of power. And the attendant cost of solar panels was at least $500 per watt.

Beginning in 1973 the groundwork for a basic industry was laid, and by 1980 solar panels could already be produced at $10 a watt. Production capacity was a few megawatts (i.e., thousands of kilowatts) per year. Direct competition with the electric utility will fully develop when the cost will be about a dollar a watt, and this figure could be achieved in the 1980s.

It is of great importance to society that alternative energy sources have net energy gains. Another idea, which I have proposed, is the creation of a solar breeder. This facility would use electricity from banks of solar cells to produce more solar cells. No outside source of power would be needed. And a solar breeder would show a net energy gain in about a year.

We have a great need for such power gains, since civilization can only flourish when an energy surplus feeds it. In this respect, photovoltaics are truly ideal. And the old debate on the merits of photovoltaic power has currently already shifted from "if" to "when" solar cells become cost-competitive with conventional power. In my view, one of society's major challenges is to maintain this positive momentum. Thus we can obtain inexpensive photovoltaic systems and realize a solar energy surplus as soon as possible.

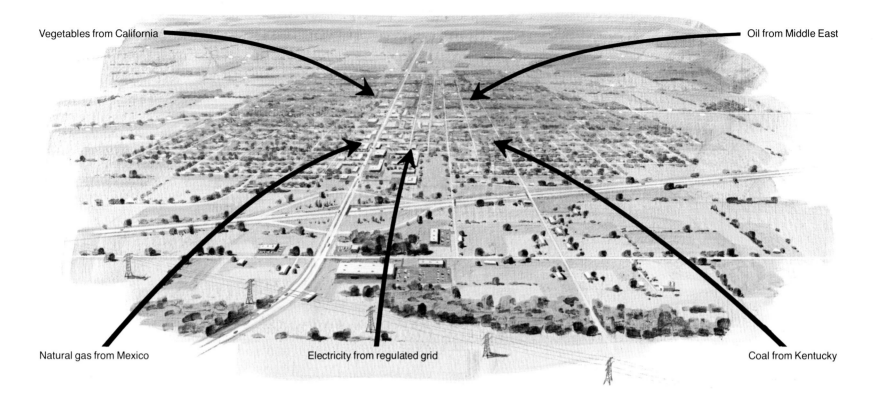

Vegetables from California

Oil from Middle East

Natural gas from Mexico

Electricity from regulated grid

Coal from Kentucky

Humanly Scaled Energy

David Morris

We began to learn that bigger is not necessarily better after one faulty piece of equipment blacked out most of the giant Northeastern power grid in 1965. Imagine, indeed, how vulnerable this nation would be if its electrical power were generated by the largest and most complex solar-electricity scheme yet conceived—a brace of immense generators orbiting in space.

By whatever route, the nation will certainly be moving into a solar age. Up until now the debate has been largely academic, centering on the issue of renewable energy sources versus nonrenewable ones. But today we are moving closer to the realm of practical planning— politics and economics. Solar society involves answers to serious questions of scale, of control, of means and ends. It is certainly none too soon to begin some hard thinking about the ways that new technology will work to restructure society.

I make the case here for the local generation of energy, where power-producing towns and farms would work in harmony with the large regional grids—a synergy of smaller and larger units without economic domination by a monolithic utility system. It is unlikely that people will be willing to pay the premium necessary to be totally self-sufficient. There are also important economic arguments in favor of systems larger than the household. I call this humanly scaled energy because, as we shall see, it directly involves the individual.

For example, MIT's Energy Laboratory matched the hourly output of energy from a rooftop photovoltaic array with the demand for energy from the household itself. The MIT engineers concluded that a 900-square-foot array could provide all household energy needs, *on an annual basis*. A great deal of the energy, however, would be generated when it was not immediately needed. The excess could be stored in batteries, but it would be more efficient to send these momentary surpluses into the grid system. In this way, a household in Phoenix or Boston would export 50 percent or more of its electrical generation and import its shortfall from the grid. I believe this kind of symbiotic relationship holds the key for future energy development.

Similar economies of scale apply for heat storage systems as well. They stem from the basic principle of physics that as we increase the volume of a tank, we do not increase its

Residential

Multi-family dwellings

Business

Industry

Gardens and orchards

} Solar arrays, thermal ponds

Electric service stations

Alcohol-electric stations

Below: proposed solar town of the future draws energy from many sources. Opposite: small urban area of the 1980s. Superimposed arrows on both indicate the place of origin of energy, food, and some materials. Note the great distances that items must be transported to supply the contemporary town. The food-distribution links of the solar city, for example, would more resemble those of the American hometown of 1890 than the conventional community of 1980. The primary difference is the marriage of high technology to local production and distribution of many of life's necessities. Solar collectors provide much of the energy.

Illustrations by Pierre Mion

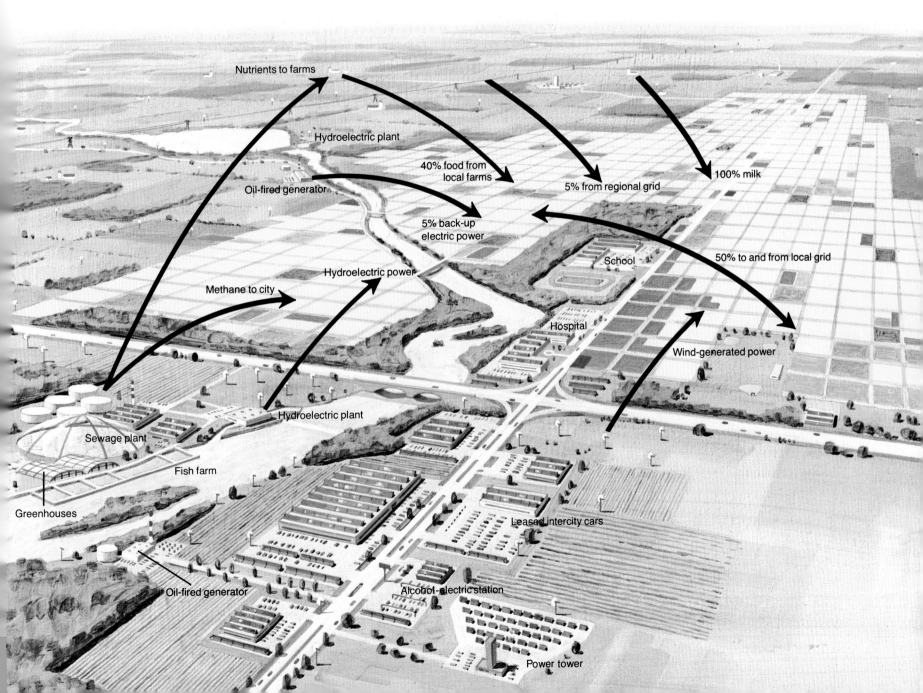

Nutrients to farms

Hydroelectric plant

40% food from local farms

100% milk

Oil-fired generator

5% from regional grid

5% back-up electric power

Hydroelectric power

School

50% to and from local grid

Methane to city

Hospital

Hydroelectric plant

Wind-generated power

Sewage plant

Fish farm

Greenhouses

Leased intercity cars

Oil-fired generator

Alcohol-electric station

Power tower

From far left: covered pond traps summer heat for storage. Distribution occurs during winter or prolonged cloudy spells. When the sun shines, houses heat themselves, even when temperatures plummet. Slanted arrays of solar cells help provide electricity—stored underground nearby. Greenhouses and close-in gardens and orchards yield produce for seasonal supply and for preserving, while regional surpluses add to winter supplies.

cost proportionately. We can double the volume without doubling the cost of construction. In Canada and Scandinavia very large storage tanks can collect the sun's energy during the long summer days and store it for use on long winter nights. One study sponsored by the Canadian National Research Council concluded that it is most economical to have thermal storage tanks serve 100 houses, assuming modest density. This system, including distribution, would be 75 percent cheaper than one in which each house had its own storage tank.

But is there enough space, especially in our dense urban areas, to generate a significant portion of our energy? Certainly no one has yet figured out a way to power the World Trade Center towers by solar energy. They consume as much electricity as the city of Schenectady. But the conventional wisdom that America is dominated by large, crowded cities is simply untrue. The United States is composed mainly of small and medium cities. In 1977 exactly half the population lived *outside* incorporated areas of more than 10,000 people. As many people live in towns of 10–25,000 residents as in cities with more than one million.

Even our larger cities have low population densities. Furthermore, most urban dwellers live in single-family homes, often with front and back yards. Backyard utilities are not so farfetched, especially with effective commercialization of existing solar energy convertors in the future.

Overall, it appears technically possible to generate a significant portion of our energy in a decentralized fashion. Only time will tell the exact shape of our energy future, but because of a piece of major legislation, the issue of decentralization is no longer theoretical. The federal government opened up the electric grid system in 1978. The Public Utility Regulatory Policies Act requires utilities to buy electricity from people who produce enough power locally to have a surplus to feed back into the grid. Utilities must offer a fair price for this electricity, while providing inexpensive back-up power.

It would be extremely speculative to describe future community energy systems in great detail. But we will certainly see the rise of concentric circles of responsibility. The household may provide 50 percent of its energy needs on-site, but may export 50 percent of the energy it produces to the neighborhood, which will provide storage systems and back-up power.

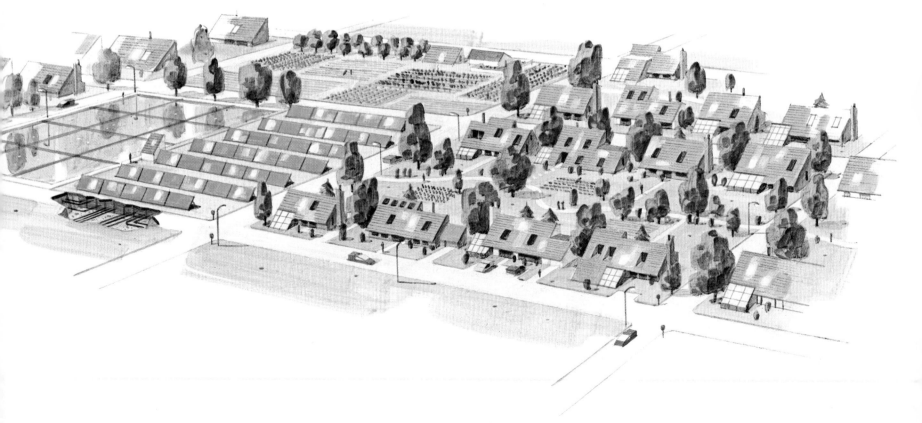

The self-reliant city will probably raise a portion of its food within the city limits. Organic wastes left over after the production of methane and alcohol fuels will be used to condition the soil or to supplement animal feed. The river running through a city will not merely be harnessed for its power, but managed and protected for its ability to produce fish. Recycling of waste would provide some jobs and become a source of metals and other materials for industry.

Solar breeders, wherein arrays of solar cells generate power to manufacture more solar cells, may well become familiar light industries, taking care of local needs and exporting excess products to neighboring towns. As the cumulative effect takes hold across one country, surplus production of cells could be exported to other lands. Large cells could become as plentiful as house shingles are today, and installed in much the same way.

Decentralized systems like the one I've described above are more stable, less vulnerable to the effects of strategic weapons, cheaper, and more efficient. They can more easily be matched to changes in demand. They can be designed to fit local climatic conditions. For

instance, depending on what part of Portland, Oregon, one lives in, the annual rainfall can vary from 27 to 64 inches a year. A six year study of St. Louis concluded that the eastern portion of the city experienced higher summer precipitation, 10 percent more cloud cover, and 30 percent more rain.

There's another important reason why small-scale can be cost-effective. A major contributor to the deteriorating economic position of major electric utilities has been their reliance on very large energy systems: huge power plants requiring 10 years to move from initial planning to first power flow. Such a long-term investment can be profitable only if one can accurately predict demand a decade into the future. That is not always possible.

In 1979 the Edison Electric Institute estimated a six percent peak electrical demand increase. The actual increase was only one-tenth as great. Idaho's energy planners determined their requirements for the mid-1990s. Their projection was made in 1979, but by the spring of 1980 other Idaho officials lowered the state's projected rate of growth from 5.5 percent to 4.5 percent. That single drop led them to subtract two billion dollars from their esti-

ELECTROHOL sign announces filling station of the future. The long distance traveler pulls in to fill his heavy car with alcohol. Long trips would take place in leased cars. Individually owned electric runabouts will be refueled simply by switching batteries. Alcohol for fuel is distilled on local farms and tanked in like milk. Solar arrays on the spot, and parabolic steam generators produce electricity for the self-contained station.

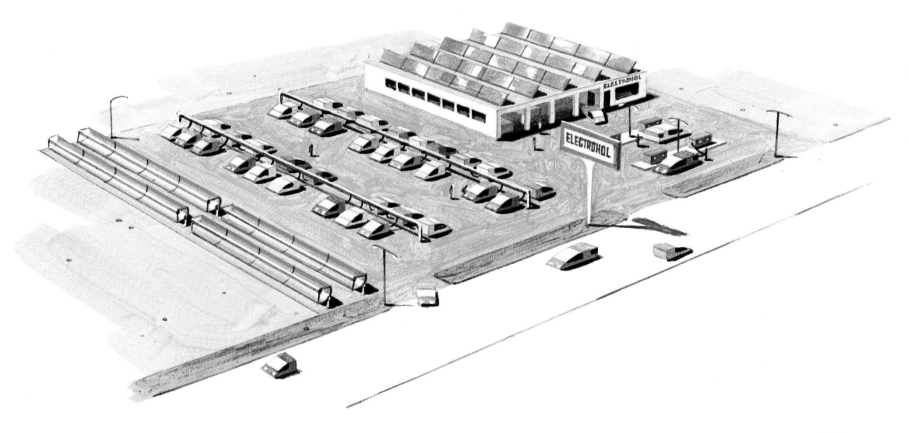

mated capital requirements. Once it's built, it's built, and must be paid for, whether the expected demand emerges or not.

Most solar technologies, being modular, offer a real flexibility. They can be built rapidly to match real demand and redeployed to support changing needs. Many solar collectors are as efficient at the household level as they would be in central applications. This is certainly the case with photovoltaics—the familiar blue solar cells. Thus they reduce our reliance on long-range forecasts which have proven increasingly suspect.

Humanly scaled power would minimize waste. For instance, the same roof surface that yields electricity from solar cells can also yield heat for both warming domestic water and providing house heat.

Large-scale systems can work against such domestic economy. In the 1930s, small electric generating plants often piped steam for heat to their customers. Today's big generating plants, though, have become so remote from customers that they can no longer deliver the heat (two thirds of their energy output) in an economic way.

The same principles of scale, economy, and conservation apply to that other form of solar energy—wind. Many modest-sized wind turbine generators appear to be more efficient than a handful of large machines. The variability of wind even within small geographic areas encourages dispersed systems. Dr. C.G. Justin of Georgia Institute of Technology says:

The beneficial effect of operating large dispersed arrays of wind turbines is that available power output can be increased—if winds are not blowing over one part of the array, chances are they will over some other part . . .

Several wind turbine manufacturers, acting on this principle, now work to establish wind "farms," clumps of medium-sized wind turbines. Woody Stoddard, an engineer for one of these companies, U.S. Windpower, sums up the technical reason why smaller wind generators are more economic:

The energy produced goes up as the square of the blade diameter (economy of scale), but the structural cost of the blade goes up as the cube of the diameter (the square cube law). Therefore you end up with machines that are so large they offer diminishing returns on the money invested. We have found that the minimum is around 50 to 75 feet in diameter, and that is the size we plan on going with.

But the primary reason that such locally scaled, decentralized systems are valuable is that they are democratic. They provide people with a power base. In 1800, nine out of ten Americans worked for themselves. By 1980, an overwhelming 95 percent of the people worked for someone else. From a nation of owners and producers, we have become a nation of employees, renters, and consumers. Humanly scaled energy systems can help people provide for themselves.

Like developing nations, communities now export a considerable portion of their income for basic goods. Local energy planning can benefit the local economy. One study of the District of Columbia found, for example, that $500 million was spent on energy of all kinds in 1980, excluding purchases by the federal government. Only 15 cents on the energy dollar returned to benefit the city in any form, whether for wages and salaries to local residents, profits or dividends to local proprietors or stockholders, or taxes to the local government. Humanly scaled energy systems can retain a greater portion of the energy dollar within the community, where it can be recycled. A study by the Institute for Local Self-Reliance of the District of Columbia concluded that for every dollar not spent on imported energy, $2.50 in gross income was generated in the community.

Furthermore, local consumer-producers can evaluate the tradeoff between such important issues as cost and reliability. Utilities currently try to maintain enough capacity to ensure that customers can obtain energy 99.999 percent of the time. One utility economist argues that this standard for reliability is too high, that the last few hundredths of a percent of reliability are enormously expensive.

Again I stress the point that while a solar society could truly put the control of energy into the hands of people, such an outcome is not inevitable. The power from our nearest star can be collected in dramatically different ways, but I believe that the different strategies for doing so are mutually exclusive. We can just drift into a system of centralized power. Decentralized energy production will require a push though, and I believe that the enduring social and economic benefits will be well worth the effort.

Power for the solar farmstead may be generated by photovoltaic cells located on south facing rooftops, and by wind turbines. Some of the corn grown in nearby fields goes to produce

Cornfield for alcohol still

Grai[n] storag[e]

Solar—methane grain dryer

Home garden

alcohol fuel for farm machinery and for sale to towns, see fermentation tank near the pond. Having produced an energy surplus, the farm turns its remaining grain into milk and meat.

Protein-rich mash from the fuel fermenting tank is added to carefully formulated rations for pigs and cows. The farmer plows composted organic wastes back into the soil, and automatic

sprinklers water and fertilize portions of fields as sensors show that water and plant food are needed.

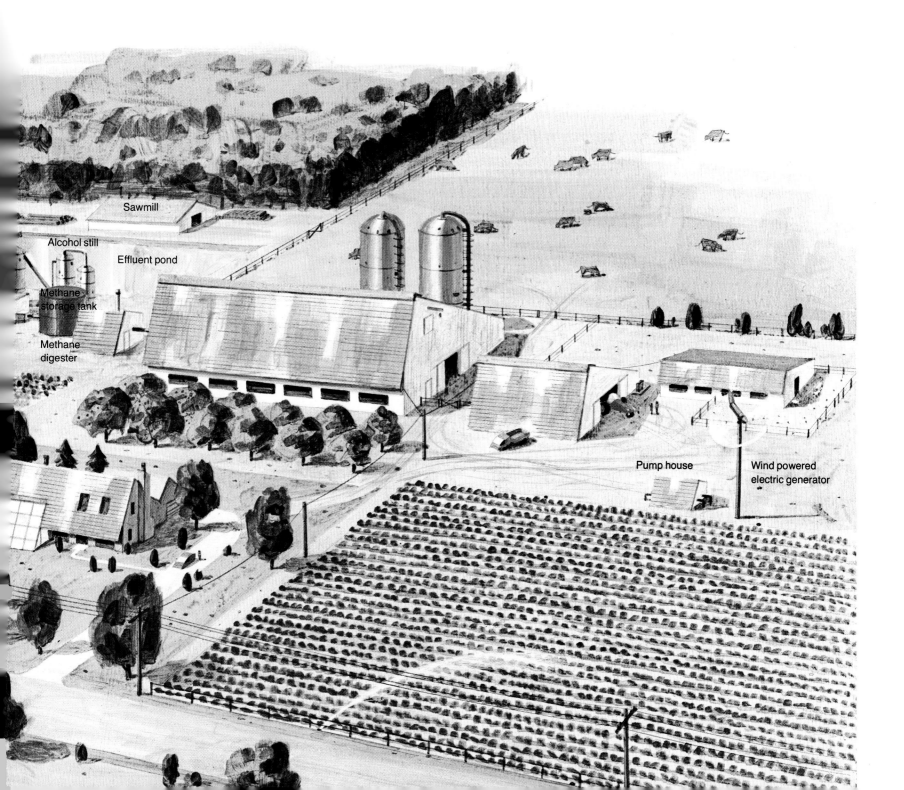

Sawmill

Alcohol still

Effluent pond

Methane storage tank

Methane digester

Pump house

Wind powered electric generator

Peter E. Glaser

Power from Earth Orbit

British physicist and molecular biologist J.D. Bernal, in his book *The World, the Flesh and the Devil*, summed up our global peril and promise in just a few words:

It may be that, in the future, man will have no use for energy and be indifferent to stars except as spectacles, but if (and this seems more probable) energy is still needed, the stars cannot be allowed to continue in their old way, but will be turned into efficient heat engines.

Bernal's statement rings even more true today, now that we have rediscovered the inexhaustible potential of solar energy. I say rediscovered because the potential of solar energy as a source of power has been recognized and tested for more than 100 years.

Efforts to harness solar energy accelerated during the last half of the 19th and the beginning of the 20th century as the world's energy needs grew to meet the demands of the Industrial Revolution. These efforts subsided with the successful development of energy economies based at first on coal and, subsequently, on the use of liquid petroleum fuels. It was not until the early 1970s that scientists and technicians began the serious development of solar energy technology that we witness today.

Solar energy, the primary source of energy for the global ecosystem, drives the hydrologic and atmospheric cycles and is the basis for photosynthesis, which sustains life in all its varied forms. It is by far the largest source of energy available to the Earth, which intercepts about 1.7×10^{14} kilowatts. The sun contributes 5000 times the total energy input from all other sources, and its availability is assured for eons to come. If tapped, it could provide virtually unlimited amounts of energy to meet all conceivable future needs. Yet today we are using

Weightless components of the first orbiting power plant seem to hang in space 22,000 miles above the planet. Vast but simple, the generator consists of a photovoltaic array and a micro-

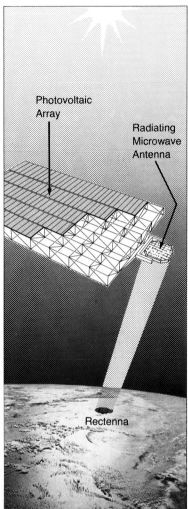

Photovoltaic Array

Radiating Microwave Antenna

Rectenna

wave antenna. Cells transform sunlight into electricity for conversion to microwaves that are transmitted to Earth. On the ground, a rectenna would intercept the microwaves and change their energy into conventional alternating current for distribution on the national utility grid.

Construction of a moon base for the extraction of minerals and manufacture of prefabricated parts might precede actual fabrication of orbiting powerhouses. Lunar ores could yield aluminium and silicon, and lifting structural elements into orbit from the moon would be easier and cheaper than launching them from Earth.

practically no solar energy. Instead, we are burning oil and gas, both of which are limited resources.

In principle, we have infinite energy in a finite world, whereas, in reality, we are using finite energy in a world that was until recently perceived to be infinite. Obviously, we cannot easily or instantly switch from our present nonrenewable energy diet to a future menu of solar energy.

There is a growing impatience to develop solar energy for widespread use, but despite optimistic expectations, the large-scale use of solar energy may take longer, be more difficult, and cost more than has been projected. What is clear is that no one energy source will meet all foreseeable future energy demands, that the search for new supplies of nonrenewable fuels can only put off the day of their ultimate exhaustion, and that there are major uncertain-

ties in achieving the potential of known energy technologies. But there is no need to rely on just one solution to the energy dilemma. The key to assuring future energy supplies will likely be the matching of diverse technologies with their most appropriate applications.

Current solar energy research and development is directed toward new and improved technology, approaches to reduce the cost of conversion, and designs and processes to permit low-cost mass production. Although expectations for significant benefits are high, we should not anticipate quick and easy results on the desired scale. The impediment today is not lack of appropriate technology, but rather lack of appreciation for the potential of solar energy and our consequent lack of experience with such technology.

Numerous studies project what will happen by the year 2000 and how to deal with global,

national, and regional energy problems. Mostly, these studies focus on changes in pricing, management, and allocation of available resources that must be anticipated within already existing economies. No matter what changes are recommended for such economies, they will not have immediate effects, and will therefore in all likelihood be resisted. Changes there must be, however, if we are to bring about the transition from oil, gas, and other nonrenewable energy resources to the renewable energy sources which will be essential to proper functioning of energy economies in the future. Therefore, the transition period will last beyond 2000. Shifting too soon or too quickly to solar energy could strain national economies. Shifting too late or too slowly might also impose inescapable pressures on some fossil fuels, resulting in sharply escalating prices and consequent damage to these economies—as happened during the Oil Embargo of 1973 and continues to this day.

Huge energy supplies will also have to become available if the developing countries are to approach the economic level of industrialized countries. The future energy requirements of developing countries are projected to be more than four times the total world energy production in 1970 (the equivalent of about seven billion tons of coal). Yet political and economic realities suggest that industrial countries will remain the major users of the world's energy resources. The prospect of supplying the equivalent of 30 billion tons of coal per year (with the resulting environmental effects) to meet the aspirations of developing countries is dismaying enough to convince me that only solar energy can contribute significantly to balanced global development.

The political repercussions of increased solar energy on a worldwide basis would be far reaching indeed and require extensive and unprecedented international cooperation. Such a movement could lead to a safer and more stable world. Can the same be said of the continued rapid exploitation of nonrenewable fuels, keeping in mind that most such fuel sources are under the political control of only a few nations favored by geographical circumstance.

Solar energy could supply a major portion of future energy needs if effective ways can be found to convert it economically and efficiently to heat, power, electricity, or to other fuels. Industries and governments of many nations are trying to find out which solar energy appli-

cations are most likely to be successful, what products should be developed, how big the solar energy market, and when the potential of solar energy will be realized on a significant scale.

The major challenge to the effective application of solar energy is that sunshine is a rather diluted energy resource, requiring large areas for conversion into useful form and, therefore, capital-intensive technologies. The successful and widespread introduction of solar energy technology will require considerable development to strike the appropriate balance among the conflicting economic, environmental, and societal considerations. But, with goodwill and good sense, such a balance can be struck.

It is my belief that solutions can be found to several of the problems of solar energy application, whether they be technical, cultural, economical, or political, if we turn our attention outward from the Earth, away from preconceptions and present-day systems, and take a thoughtful look at the possibilities space offers.

The launching of Sputnik on October 4, 1957, and the subsequent dramatic unmanned and manned space pioneering efforts marked our entry into the Space Age. These events irrevocably changed the evolutionary directions of Earth's civilization and brought to us a consciousness of the uniqueness of our planet. At the same time, many people began to recognize the great powers and advantages inherent in our kit of Space Age tools. Rockets, computers, and communication systems not only promised an unlimited extension of new knowledge of the solar system, but had a most profound influence on advances in technology.

The historic "one small step for man, one giant leap for mankind," taken in July 1969, has very practical long-range connotations: just as the railroad in the 19th century opened up frontiers for human settlement, so space transportation sets the stage for the movement of humanity beyond the Earth's surface. The railroad was a technological solution to social problems: unsatisfactory social conditions caused by resource limitations were alleviated when the railroad facilitated the exchange of the agricultural products and natural resources of new settlements for the manufactured goods of overcrowded cities. Similarly, space transportation can bring many benefits within the reach of Earth's civilization.

In the 1960s, the logical soundness of marrying solar energy conversion technology and space technology led to the concept of the solar

Realization of the author's concept for solar power satellites depends in large part on perfecting Space Shuttle operations as well as the early development of even larger freighters for use between Earth and the moon.

power satellite (SPS). The SPS would be placed into geosynchronous Earth orbit. (An object in geosynchronous orbit, about 22,500 miles from Earth, takes the same amount of time to complete one orbit as the Earth does to rotate once on its axis. Thus an object placed in orbit at this distance will remain over the same point on the Earth's surface.) There it would convert solar energy into electricity and feed it to microwave generators forming part of a transmitting antenna. The antenna would precisely direct a microwave beam of very low power to one or more receiving antennas at desired locations on Earth. On the ground, the microwave energy would be safely and efficiently reconverted to electricity and then transmitted to users. An SPS system would consist of a number of satellites in Earth orbit, each beaming power to one or more receiving antennas.

At the outset, the SPS concept challenged the view that solar energy conversion could not make a significant contribution to energy economies, and demonstrated that there are no *a priori* limits to the development of energy resources in space. Although not a panacea, the SPS could open up a new evolutionary direction for the human development of energy resources.

Today, we have shed the illusion that we have unlimited capabilities to control and fashion our environment. And although we have not yet used up the resources available on Earth, the dynamic growth which was built on a century-long exploitation of cheap and abundant energy supplies appears to be at an end. Within our grasp we not only could have the energy available in space, but we would not even have to rely on terrestrial resources to build up an industrial capability in space. There is increasing confidence that with presently known space technology, we could take the elements in moon rocks—oxygen, silicon, and aluminum—and convert them into construction materials for an SPS. Even industrial complexes no longer would need to depend exclusively on Earth's resources to meet terrestrial needs.

As originally conceived, an SPS can use today's well-understood approaches to solar energy conversion, that is, photovoltaic and thermal-electric, as well as others likely to be developed in the future. Photovoltaic conversion represents a useful starting point because solar cells are already in wide use in satellites. An added incentive is the substantial progress being made in the development of low-cost, reliable photovoltaic systems, and the increasing confidence in the possibility of achieving the required production volume (see pages 222–229).

Because the photovoltaic process is passive, it could give the SPS an operating lifetime of at least 30 years and, with maintenance, perhaps even several hundred years. Micrometeoroid impacts are projected to degrade one percent of the solar cell array area over 30 years, with only small probability of impact by larger bodies. And repairing this minimal meteorite damage would be virtually the entire maintenance task.

To transmit the power generated in the SPS to Earth, either a microwave beam or a laser beam system could be employed. Transmission of power by microwaves through space already has been successfully demonstrated and new technology may eventually raise the low initial efficiency of 55 percent up to 70 percent.

The transmitting antenna for the SPS reference system is designed as a flat disk having a diameter of about one kilometer (0.6 mile). Back on Earth the receiving antenna should be about ten kilometers (six miles) in diameter. The power density will be highest at the middle of the receiving antenna, and will decrease with distance from the center. The receiving antenna's collection efficiency is insensitive to substantial changes in the direction of the incoming microwave beam; efficiency is not affected by most atmospheric conditions. More important, the process of reconverting the microwave beam back into electricity creates less thermal pollution than any known thermodynamic conversion. The receiving antenna could also be designed to be partially transparent, so the area underneath it could be put to other uses, including specialized farming or even grazing. Floating offshore receiving antennas are of particular interest because of limited land availability near major urban centers and the frequent proximity of these centers to coastal locations.

Microwave power transmission is the present choice, based on considerations of technical feasibility, fail-safe design, and low energy flux levels. Laser power transmission is an interesting alternative, however, enabling engineers to deliver power in quantities as small as 10 megawatts to individual receiving sites on Earth. There, mercury-cadmium-telluride photovoltaic cells could be used to convert the energy in carbon dioxide laser beams from the SPS directly into electric power.

Once solar power satellites and their Earth stations reach full operating capacity, nearly all of a nation's electricity could come from the sun, with products from Earth's fossil fuels used as chemical feedstocks.

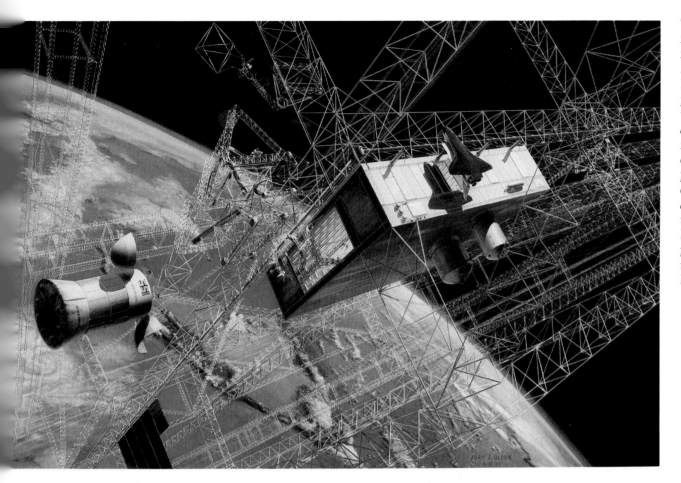

The weightless, airless environment of space will foster the new field of orbital engineering. Spidery structural elements—beams made of wire and tempered metal foil—could tether vast solar arrays weighing hundreds of tons which would require massive supports if built on Earth. Delivery of the right parts to the proper technicians on schedule would require vast expenditures on an international basis during decades of base construction and in the final years when the satellites themselves are fabricated and brought into service.

Whatever the orbital equipment and the earthside setup, the SPS will require a space transportation system capable of placing payloads into geosynchronous orbit at low cost. The Space Shuttle, now well along in development, should blaze this economic trail. Compared to the previously used expendable launch vehicles, the Shuttle will not only significantly reduce the cost of launching payloads, but will also be a major step toward the development of true space freighters of greatly increased payload and substantially lower costs.

As development of space transportation systems proceeds, costs should drop during the 1980s from the thousands of dollars per pound launched during the Apollo moon missions to hundreds of dollars per pound for the Shuttle. Space freighters of the more distant future will boost cargo for tens of dollars a pound.

Zero-gravity conditions present a unique freedom for the actual assembly of structural materials delivered to the site. But new and unusual constraints can be expected, also. The structure of the planned SPS will be larger than any ever fabricated on Earth. Therefore, innovative construction methods will be required for the structures which will hold major SPS components. These include modular solar arrays which will be linked to form the great solar collectors, and microwave subarrays to form the transmitting antenna. The structure's immensity ensures that it will shrink and expand considerably as a result of large changes in temperature imposed on it during short eclipses around the time of the equinoxes as it orbits the Earth, passing from sunlight to shadow and back again. During such eclipses, temperature variations as large as 200° Kelvin (400°F) could be imposed. Depending upon the SPS's dimensions, changes of 50 to 100 meters overall could result if a relatively inexpensive aluminum alloy is used. The more expensive graphite composites also under consideration are not particularly subject to thermal expansion or contraction; however, the aluminum structure could be insulated.

While such technical problems are under study, economics too must be weighed. Justifi-

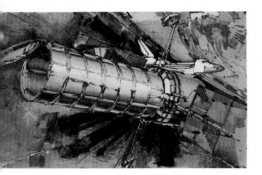

Advances in space telescope technology would be just one spinoff from the vast solar satellite project. Manufacturing electronic components, pharmaceuticals, and specialty alloys would all be made more efficient in the weightlessness of space.

cation for proceeding with an SPS development program is based on a classic balance of risks against benefits. We must acknowledge, while conducting this analysis, that we cannot really be sure of the cost of technology which will not be fully developed for at least 10 years. (And commercial exploitation of the system is at least 20 years away). Justification, of course, is equally difficult to provide for other advanced energy projects. It becomes apparent that this approach, therefore, requires an appreciation of the competitive cost of alternative energy sources for the generation of electrical power which would be available in the same period.

Any SPS development program should be time phased. Thus the "economic" purpose of each program segment will be to obtain information that will lead the decision makers to a deliberate decision to continue the program or to terminate it—and thereby control the risk.

Cost-effectiveness analyses alone would be inappropriate as they would require very long-term projection of future needs—and this is a risky art indeed. The near-term decisions regarding the SPS program should be based on resources allocated to SPS-related research tasks and their priorities rather than to the projected economics of the SPS.

On the basis of present value, a cost analysis of supplying nonrenewable fuels such as coal to power-generating plants, compared with the cost of the SPS, demonstrates that SPS costs are within the competitive range. Such comparisons of different energy technologies, however, are complicated unless external costs, particularly those due to environmental impacts and societal effects, are appropriately recognized and included as part of the total cost of a specific energy technology.

Other less obvious costs must be accounted for as well. And the SPS is unique because for the first time a major program is being evaluated on the basis of the full spectrum of society's concerns. These include environmental effects, comparative economic factors, impact on existing institutions, as well as program risks and uncertainties. Such a comprehensive approach is especially appropriate at this time of public skepticism about complex, large-scale technologies.

Effects on human health and the total ecological impact of microwave power transmission are key environmental issues. And indeed large populations may be exposed to very low levels of microwave energy. Investigations indicate

that workers when properly protected could safely maintain the SPS in orbit and the receiving antenna on Earth even while these are in operation. According to our present scientific evidence the public beyond the receiving antenna site is safe. Nevertheless, we must continue to conduct research on microwaves and their biological effects to assure that life-long exposure to a low background of microwave energy is, in fact, acceptable.

In addition, liquid hydrogen and liquid oxygen, which are the most likely propellants for the space transportation system, may interact with the upper atmosphere. For example, they could change the characteristics of the Van Allen radiation belt, which in turn may affect satellite communications. They could slightly decrease the concentration of ozone in locally limited areas of the ozone layer. The launch, however, could be controlled to minimize such effects by controlling the booster rocket to reduce the rocket emissions during transit through the layer. Rocket exhaust could locally change the properties of the ionosphere, interfering with that layer's signal-reflecting properties now exploited in certain telecommunications. Such rocket exhaust interactions with the upper atmosphere eventually could be nearly eliminated by using extraterrestrial materials for the construction of the SPS.

The microwave transmission from the SPS also poses a potential source of direct electromagnetic interference, with radio for example. But it is quite difficult to figure just what might be the impact of this factor in the year 2000. Radio and optical astronomical observations might also be significantly inhibited by microwave transmission and by light reflected from the surfaces of a large number of SPSs which would form a global system. But there are some splendid technical solutions available for just such problems—orbiting astronomical observatories, for instance.

Obviously, any environmental effects induced by the SPS have to be compared with the environmental effects of alternative power-generation methods. A detailed analysis is beyond the limited scope of this essay. Suffice it to say that an analytical framework is being formulated—one that may help to swing the balance in favor of SPS.

Basic comparisons do indeed indicate that the SPS has the potential to be environmentally benign, even if introduced on a global scale. But continuing investigations are necessary to reduce the uncertainty of any environmental risks associated with the SPS, especially in view of its potential to be a power-generation method of global benefit.

Whatever the technical considerations, they will—and certainly should—be viewed in the light of their potential impact on society. Fundamental considerations include the ownership of the SPS, the responsibility of the owners in case of accidents, and the potential vulnerability of the SPS to actions of adversaries. New institutions may have to be created to construct and operate large centralized energy-production technologies so that they can be controlled by the societies they serve.

The views and opinions expressed by those in support of or in opposition to the SPS concept represent widely different attitudes and ideological beliefs. The contributions of distributed (dispersed) and centralized technologies, accountability of industry and government, participatory democracy, the price and availability of nonrenewable fuels, the environmental and health impacts of alternative energy technologies, and the degree of international cooperation will influence the future course of development of the SPS.

The SPS appears to involve technologies which are at opposite ends of the scale of dispersed terrestrial solar technologies, the kind called "decentralized" or "humanly scaled" in Dr. Morris's essay preceding mine. In my opinion, the differentiation of solar technologies according to the sizes of their conversion and distribution systems introduces an artificial barrier which may hinder rather than advance the development of the most appropriate solar technologies to meet eventual requirements.

Solar cells, with only minor modifications, could be used on Earth or in space. And curiously, SPS development could encourage distributed solar technology because photovoltaic research for the SPS could also be beneficial for terrestrial photovoltaic applications. Most likely, I suspect, strong central planning will be required for both dispersed and centralized solar technology applications to succeed. Awareness of these technologies and their potential by individuals, communities, regions, and countries will surely differ at various stages of technology development.

The SPS could provide not only the impetus for peaceful cooperation among nations because all can share the limitless resources of space, but it could help us achieve the inevitable transition to renewable sources of energy, inevitable, that is, if our advancing world civilization is to endure and mature.

By the way of conclusion, I affirm—along with the editors of *Fire of Life*—that we are approaching a fork in the energy resource road. I believe that we need to choose most carefully the direction we take. Up to now we have tended to look for solutions of our immediate problems only, and we have often failed to see the future except as a continuation of the present. Perhaps we do not care about the distant future because we will not be there to enjoy it, and we are willing to let the next generation deal with the problems we bequeath to them.

I believe that society can meet its basic responsibility for building a livable future for succeeding generations, and that all men can prosper in the process. The key is the use of renewable resources; sun power represents the ultimate energy resource.

Planetary pivot shot— sun to satellite to Earth— might well provide a key to future international cooperation. Plentiful energy for all nations could help break down barriers to international accord that have plagued mankind for centuries. The Intelsat union for international satellite communications provides a contemporary example of just such an association.

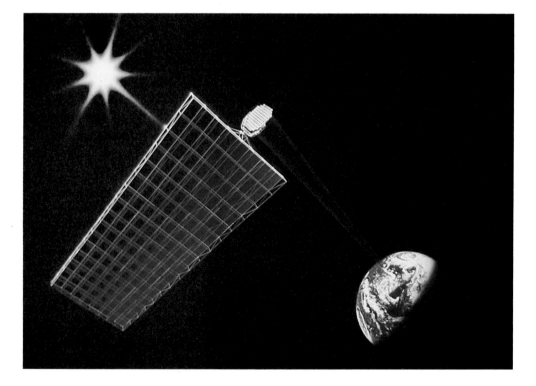

Von Del Chamberlain

Epilogue

A Song for the Sun

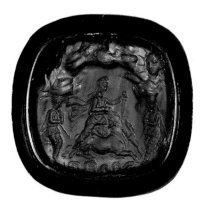

Dimmed by distance, all above and around me are the night suns which I, as an astronomer, spend so much time thinking about. Our own sun is merely one of these. If we were out there looking back, the sun would not impress us in any particular way. Actually, it is a dim, rather dull type of star. Yet what has occurred on this planetary member of the star system is anything but lackluster. Life is varied and vigorous.

My quest, my own musing, brings me to a hilltop in the predawn hour. I have come early so that I might witness the entire drama of the emergence of our daystar.

Over there, to my left, is the star known for the fact that it does not appear to move, the North Star. It is much bigger and brighter than the sun, it changes brightness periodically, and it has a dim companion star.

Across the sky, to the south, stands Antares, distinguishable by its red color and its association with the fishhook-shaped group of stars in the constellation Scorpius. If this supergiant star were placed where the sun is now, it would engulf the orbits of all the inner planets including Earth and Mars.

As my eyes trace upward along the rim of our Milky Way, I recall the words, "the prairie is dark, but across the sky is a trail of light. It is the ghost pathway of the departed warriors." This Otoe Indian saying suddenly acquires a different meaning than that intended. How many "warriors" have lived and departed out there among those stars which blend to form this milky splash across our sky?

My eyes come to rest on a faint star in the constellation of Cassiopeia. I know that this is one of the few stars we can pick out with our unaided eyes which has the same physical properties as the sun, our own special star. It is yellow like the sun and has the same temperature and size. It is, however, different in one important aspect. It has a dimmer, cooler, smaller star moving with it.

I like to imagine that both of these stars, so near to each other, have planets with intelligent life on them. How interesting it would be when they discovered each other. They might share their histories, even blend their cosmically different cultures.

So many things are possible out there because of the stars. Modern astrophysics has shown that stars are one of nature's most profound entities. They first gained shape and energy by gathering in the simplest state of matter—gaseous hydrogen and helium. But once they burned, stars helped to build a far more complex universe through thermonuclear fusion. This same process enriched the universe with new, heavier elements.

Eventually, it became possible for planets to exist; even planets with oceans, lakes and streams, rocks and soil, flowers and trees, birds and animals, and people. And many stars out there which send their light into my eyes could have planets with hilltops where creatures sit at this very moment, waiting for their sun to flood the sky with energy. I am aware of such

Below: sacred to the Greek Island of Rhodes, sun god Helios daily drives his fiery chariot across the sky. Farther east, from the Iranian Plateau, Sol Invictus or the "Unconquered Sun" dawned for followers of the Mithras mystery cult. A Roman lapidary engraved the gemstone, opposite, that reveals a legendary bull sacrifice replete with Mithraic symbolism of a mythic cave, serpent, and the outpouring of the beast's life-blood. Mithraism proved especially popular with soldiers in Rome's legions, and the faith rivaled early Christianity and spread throughout Europe and the Near East. Right: astride a

great lion made of mingled people and animals, a Hindu goddess holds the sun in its splendor. At first glance, this and the varied graphic idioms that follow might suggest scenes from other planets. The emotional content, though, bears witness to a universal sharing of great wonder and mystery implicit in the Earth's relationship to the daystar. Indeed, the author speculates, stellar and solar concerns of creatures on other planets may parallel those of Earth's inhabitants.

Golden sunburst and halo cast beatific auras upon mother and child in holy images from different eras and traditions. Right: surrounded by scenes from Christian iconography the Madonna, "Holy Mary, Mother of God," sits holding the infant Jesus. The Greek Orthodox icon was painted by Theodore Paulakas, an artist working on the island of Crete in the 17th century. Opposite: Hittite neck pendant of shaped and soldered gold from about 1200 B.C., represents a goddess of Anatolia. She has been tentatively identified as the Sun Goddess of Arinna, "Queen of the Land of Matti, Queen of Heaven and Earth." The child may be a weather god, or the dawn sun itself as in similar Egyptian Isis figures. In Near Eastern and North African mythology, the infant grew up within the day, dying as an old man at sunset.

possibilities because we understand that the sun is a star. That bit of knowledge is so fundamental to the way we regard ourselves in the universe. How ironic that we cannot attribute the discovery to a particular individual, time, and place. The concept crept up on us through a series of other discoveries, and it caused us to face new levels of reality. And perhaps the most interesting realization is that we are all clumps of recycled stardust.

To the west the sky is still dark, but the eastern stars are fading now. A curtain of white light moves upward and a period of innumerable changes begins. The occasional call of a bird is joined by distant tones of other birds, the sound becoming rapidly more complex. The music grows and modulates as the changes in the world about me become too rapid for my limited consciousness to enjoy. The white light of dawn is washing the canvas of the sky, preparing it for the colors which will soon follow. Low on the horizon a touch of orange is just perceptible. I make my first prediction of where the sun will cut the horizon.

Now the horizon is glowing as if on fire. I am reminded of the Skidi Band Pawnee and their concept that the Morning Star stands on a bed

247

Colorful parasol display cascades in the atrium of a Tokyo department store. Below: Japanese thunder demon rages at an umbrella carrier. Terminology varies with use, umbrella denoting rain gear while the word parasol connotes strolls on sunny summer afternoons.

of hot flint out of which the sun derives its brilliance. For a moment I feel a special closeness to those vanished people of the North American prairie. The sound of an automobile brings me back to the present, and I notice a few other man-made noises encroaching upon the melodic tones of nature.

It occurs to me that all these sounds, both natural and man-made, are solar sounds. They are expressions of energy which are actually transformed power from the sun. The pleasant perfume of morning moisture presents me with still another new thought. We can smell the sun. Indeed all odors are but special manifestations of solar energy.

Abruptly all my previous thoughts vanish as the solar disk makes its appearance in a glorious burst of brilliance. The volume of sound and color, of energy expressed in one tiny moment, is unmatched. Momentarily I feel an identity with all of humanity, aware of the significance of what we see as the sun rises. I am Egyptian, the radiance of Re flashes across the living waters of the Nile before me; I am Greek, the golden chariot of Helios rushes forward through the crimson portals of the day; I am American Indian sun watcher, taking note of the exact location of the house from which the sun emerges; I am modern American, anxiously desiring to harness the energy which flows over me. The globe moves upward,

Amaterasu, a sun goddess of popular Japanese lore, emerges from her cave to shine on admirers who have grown weary of the darkness during a time their dazzling heroine hid. The trouble began after the goddess, the Great-August-Spirit-Shining-in-Heaven, was offended by her brother, Susano-O-no-Mikoto, a storm god (see illustration opposite). She hid in her cave while legions of heavenly spirits assembled outside to make a great noise and display of jewels to entice her out. One female spirit so delighted the assembly with her dance that the gods roared with laughter. Amaterasu grew curious and looked out. The gods then held up a mirror to the face of the goddess, and she was so taken by her own bright reflection that she once again bestowed her light upon the world. The allegorical account of this Shinto sun goddess roughly parallels the sequence of events during a thunderstorm.

changing to intense orange, then yellow, and finally white. I must be careful not to look directly at it.

The sunrise tide has swept past me now. Somewhere to the west it is just arriving and still further to the west it approaches a land that rests in shadow. Except where planetary bodies make such shadows, all of solar space is pervaded by intense energy. The Earth has turned me from such a shady place toward the stellar hearth which kindles our lives.

Now that I stand in the full intensity of the sun, I notice things which mar the feelings that I have been enjoying. Many others have been here and left reminders of their presence in the form of beverage cans, broken bottles, scraps of paper, and other trash. The spell is broken.

Throughout the daylight hours of this Sunday filled with stellar musings, I think about the significance of our closest star.

As darkness approaches I return to the hilltop once more. The light fades and the birds are singing again, but I cannot enjoy them as I had just a few hours before in this same place. Now the voices from the trees are drowned out by the sounds of cars moving along the streets.

I stand but a moment on the top of the world and close out the noises and the trash around me. I think of the ancient American who might have stood here at close of day centuries ago to utter the hope of all humankind. I span the time between us and join him in his prayer:

Thank you sun.
Thank you for your warmth and light.
Have mercy on us human beings.
Come again tomorrow.

Balkan mower fells hay in a spiral design found in many lands and often associated with the soil's fecundity, human sexuality, and the fertilizing power of the sun on earth and sea. *Above:* South Seas woman quietly, privately celebrates the sunset. *First Overleaf:* Egyptian village family looks up from the morning hearth to welcome early light. *Second:* Wyoming sun momentarily silhouettes elk.

The Authors

Edward S. Ayensu, Director of the Office of Biological Conservation of the Smithsonian Institution, has contributed extensively to the literature of tropical biology, especially on endangered plant species. His wide field experience in the tropics provided the background for editorship of the recent book, *Jungles.*

Silvio A. Bedini, Keeper of the Rare Books at the Smithsonian, is a historian of science, specializing in the history of scientific instruments and of the mathematical practitioner. His works include *Thinkers and Tinkers: Early American Men of Science.*

Robert Carola writes basic science texts for American colleges and universities. His most recent such work, *Biology,* is currently in its third printing.

Von Del Chamberlain is the Astronomer at the Smithsonian's National Air and Space Museum, where he also researches Native American ethnoastronomy.

Wilson Clark, currently a consultant on energy and technology, served from 1976 to 1979 as chief energy adviser to California Governor Edmund G. Brown, Jr. He is an adviser to the municipal government of Peking on energy modernization. He has written a major government study of the relationships between energy policy, decentralization, and war.

Tom D. Crouch is a Curator in the Aeronautics Department of the National Air and Space Museum. His works on aerospace history include *Charles A. Lindbergh: An American Life.*

John A. Eddy, Senior Scientist at the High Altitude Observatory of the National Center for Atmospheric Research in Boulder, Colorado, serves also as Research Associate of the Harvard-Smithsonian Center for Astrophysics. His recent book, *A New Sun,* describes in popular terms our new knowledge of the sun obtained from Skylab observations and other recent sources.

George Field, a theoretical astrophysicist, is Director of the Harvard-Smithsonian Center for Astrophysics. In this position he helps to direct and coordinate the activities of the Smithsonian Astrophysical Observatory and the Harvard College Observatory. He is co-author of *Cosmic Evolution,* an introductory text.

Valerie J. Fletcher is Assistant Curator for Research at the Smithsonian's Hirshhorn Museum and Sculpture Garden, specializing in 19th- and 20th-century painting and sculpture. She lectures on Impressionism and Post-Impressionism for the Smithsonian Resident Associate Program.

Kenneth L. Franklin is Staff Astronomer with the American Museum-Hayden Planetarium. He has been associated with the research and educational activities of the Hayden since 1956. Dr. Franklin was co-discoverer of radio emissions from Jupiter.

Alistair B. Fraser, Professor of Meteorology at Pennsylvania State University, began avidly photographing the sky when working as a weather forecaster in Vancouver, Canada. He is a recognized authority on visual phenomena, and author of articles in this field.

Kendrick Frazier, Editor of *The Skeptical Inquirer,* is a well-known science writer whose first book, *The Violent Face of Nature,* was published in 1979.

Owen Gingerich is an Astrophysicist at the Harvard-Smithsonian Center for Astrophysics and Professor of Astronomy and the History of Science at Harvard University. An authority on Renaissance astronomy, he has specialized in the work of Copernicus and Kepler.

Peter E. Glaser, a Vice President of Arthur D. Little, Inc., also serves as President of the Sunsat Energy Council and as a Director of the American Astronautical Society.

Evan Hadingham, a native of England, is the author of several popular books on archaeology, including *Circles and Standing Stones*—which interprets Stonehenge and the mysterious megaliths of northern Europe. He has recently moved to America and is writing a book on prehistoric astronomical sites.

Sterling B. Hendricks, one of the discoverers of the plant pigment that controls responses of plants to the length of the day, is a collaborator with the Environmental Quality Institute of the U.S. Department of Agriculture. He is recipient of the National Medal of Science.

Edwin Kiester, Jr., a California writer, specializes in medical and scientific subjects. He is Editor of the *Better Homes and Gardens New Family Medical Guide.*

E.C. Krupp, an astronomer, is Director of Griffith Observatory in Los Angeles. An active participant in the interdisciplinary study of archeoastronomy, his publications include *In Search of Ancient Astronomies.*

Joseph Lindmayer, a pioneer in the field of photovoltaics, founded and now serves as President of the Solarex Corporation in Rockville, Maryland.

William C. Livingston, as Astronomer at Kitt Peak National Observatory in Arizona almost from its inception, developed instruments for the McMath telescope to measure solar magnetism. He is currently involved in helping China and India acquire modern solar facilities.

John E. McCosker, Director of the Steinhart Aquarium of the California Academy of Sciences, is also adjunct professor of marine biology at San Francisco State University. His extensive research activities have included the biology of bioluminescent fishes and the fishes of the Galápagos Islands.

Pierre Mion, a freelance illustrator and photographer, specializes in drawing, painting, and photographing scientific and engineering subjects.

David Morris, Director of the Institute for Local Self-Reliance in Washington, D.C., wrote *Neighborhood Power: The New Localism* and other books and articles on decentralized planning.

David Morrison, Professor of Astronomy at the University of Hawaii, specializes in studies of the planets and smaller bodies of the solar system. In addition to work with ground-based telescopes, he participated in the Mariner 10 and Voyager planetary missions.

George Paulikas, an expert in radiation belt physics, is Director of the Space Sciences Laboratory of The Aerospace Corporation in El Segundo, California.

Cyril Ponnamperuma is Professor of Chemistry and Director of the Laboratory of Chemical Evolution at the University of Maryland. He has written over 200 publications related to chemical evolution and the origin of life.

S. Dillon Ripley, noted ornithologist, ecologist, and conservationist, is the eighth Secretary of the Smithsonian Institution.

Illustrators

Larry Bowring—Bowring Cartographic Research and Design, Arlington, Virginia

Jackie Leatherbury Douglass and John Douglass—Parrish Creek Artists, Inc., Shadyside, Maryland

John Huehnergarth—Princeton, New Jersey

Jim Lamb—Burbank, California

Charles Murphy—Charles Murphy, Inc., Arlington, Virginia

Helmut K. Wimmer—Art Coordinator, Hayden Planetarium, New York, New York

Special Thanks

Ronald Angione—Department of Astronomy, San Diego State University, San Diego, California

Walter Angst—Conservator, Conservation Analytical Laboratory, Smithsonian Institution

James C. Cornell, Jr.—Publications Manager, Harvard-Smithsonian Center for Astrophysics, Cambridge, Massachusetts

Robert A. DeFilips—Museum Specialist with the Department of Botany, National Museum of Natural History

William A. Deiss—Deputy Archivist, Smithsonian Institution Archives

Farouk El-Baz—Research Director, Center for Earth and Planetary Studies, National Air and Space Museum

William H. Evans—Index, Rockville, Maryland

Jon P. Freshouer—Registrar, Exhibits Branch, Library of Congress

Beth Gantt—Biologist, Radiation Biology Laboratory, Rockville, Maryland

William A. Good—Museum Specialist, National Air and Space Museum

James M. Goode—Curator, Smithsonian Institution "Castle" Building

Walter Herdeg—Editor of *The Sun in Art*, 1962, The Graphis Press, Zurich, Switzerland

Leo J. Hickey—Curator, Department of Paleobiology, National Museum of Natural History

Richard Hofmeister—Chief of the Special Assignments Branch, Office of Printing and Photographic Services, Smithsonian Institution

Nicholas Hotton III—Curator, Department of Paleobiology, National Museum of Natural History

William H. Klein—Director, Radiation Biology Laboratory, Rockville, Maryland

Jack F. Marquardt—Chief of the Central Information Services, Main Library, Smithsonian Institution

William G. Melson—Curator, Department of Mineral Sciences, National Museum of Natural History

Diana Menkes—Editor/Proofreader, Washington, D.C.

Bruce Needham—Chief, Data Services Branch, National Oceanic and

Atmospheric Administration/ Environmental Data and Information Service

Nicholas Panagakos—Public Affairs Officer, Office of Space Science, National Aeronautics and Space Administration

Agnes Paulson and Connie Rodriguez— Public Information Officers, Kitt Peak National Observatory, Kitt Peak, Arizona

Robert Roosen—Department of Astronomy, San Diego State University, San Diego

Mary A. Rosenfeld—Assistant Special Collections Librarian, Smithsonian Institution Libraries

Henry Rosenthal—Rhina Color Services, Ltd., New York, New York

Glenn Sandlin—Naval Research Laboratory, Washington, D.C.

Janette Saquet—Librarian, The John Wesley Powell Library of Anthropology, Smithsonian Institution Libraries

Jurrie van der Woude—Public Information Officer, Jet Propulsion Laboratory, Pasadena, California

William Waller—Technical writer and editor, Harvard-Smithsonian Center for Astrophysics, Cambridge, Massachusetts

John P. Wiley, Jr.—Editor, *Smithsonian* Magazine

Ray Williamson—Project Manager, Office of Technology Assessment, Washington, D.C.

Harold Zirin—Professor of Astrophysics, California Institute of Technology, Pasadena, California

Picture Credits

Jacket: James Tallon. *Front Matter:* p. 1 Palomar Observatory, California Institute of Technology; 2–3 Ray E. Ellis/Photo Researchers, Inc; 4–5 Warren Bolster/© Surfer Magazine; 6–7 Kay Chernush; 8–9 Loren McIntyre; 10–11 High Altitude Observatory, National Center for Atmospheric Research. *Introduction:* p. 12 Charles H. Phillips; 13 Eric Long; 14 (top) Museum of the American Indian, Heye Foundation, N.Y., photo by Carmelo Guadagno; (bottom) Wayne Davis; 15 Kjell B. Sandved; 16 Gary Ladd; 17 Yoichi Okamoto.

Section 1: p. 18–19 Flip Schulke/Black Star; 21 Sandak Inc.; 22–23 Jim Lamb; 23 (top) *The Costumes of the Original Inhabitants of the British Islands,* 1815, by Charles H. Smith, George Peabody Dept. of Enoch Pratt Free Library, Baltimore; 24 Courtesy of Danish National Museum, photo by Lennart Larsen; 25 Anthony Weir/Janet & Colin Bord; 26 (top) Von Del Chamberlain; (bottom) Rodney Bond/ Adespoton Film Services; 27 Von Del Chamberlain; 28–29 Victor Englebert/ Photo Researchers, Inc.; 30 Robin Rector Krupp; 31 Brian Brake/RAPHO, Photo Researchers Inc.; 32 (left) Special Collections Branch, Smithsonian Institution Libraries; 32–33 Walters Art Gallery, Baltimore, photo by Charles H. Phillips; 33 (left) Erich Lessing/Magnum Photos Inc.; (right) Harrison D. Horblit Collection, photo by Owen Gingerich; 34 Derby Museums & Art Gallery, Britain; 35 Erich Lessing/Magnum Photos Inc.; 36 Jonathan Blair/Woodfin Camp, Inc.; 37 R.H. Sanders & G.T. Wrixon, National Radio Astronomy Observatory; 38 *American Journal of Science and Arts,* Series III, Vol. 9, 1875, Smithsonian Institution Libraries, photo by Charles H. Phillips; 39 Smithsonian Institution Archives; 40 *Professional Papers of the Signal Services,* 1884, Smithsonian Institution Libraries, photo by Charles H. Phillips; 41 Smithsonian Institution Archives; 42–43 Ted Hardin; 44 Naval Research Labora-tory; 46 (top left) Phil Jordan; (top center) NASA; (top right) Marshall Space Flight Center; (center) American Science & Engineering/Harvard College Observatory; (bottom) High Altitude Observatory/ Marshall Space Flight Center; 47 (top left, right & center) © AURA Inc., Kitt Peak National Observatory; (bottom left) Harvard College Observatory; (bottom right) Naval Research Laboratory; 48 (left) © AURA Inc., Sacramento Peak Observatory; (right) Radiation Biology Laboratory, Smithsonian Institution, photo by Charles H. Phillips; 50–51 Art by Jackie Leatherbury Douglass, photo by Charles H. Phillips; 52 © AURA Inc., Kitt Peak National Observatory; 52–53 Gary Ladd; 53 Modified version of the McMath Solar Telescope at Kitt Peak National Observatory, as adapted by Bowring Cartographic Research & Design.

Section II: p. 54–55 Jonathan Blair/ Woodfin Camp, Inc; 56–58 © Palomar Observatory; 59 © AURA Inc., Kitt Peak National Observatory; 60 Jim Lamb; 61 Department of Astronomy, University of Michigan; 63 From *Rose Windows,* Painton Cowen, Thames and Hudson Ltd., London; 64–71 Helmut K. Wimmer; 72 (top) NASA /Jet Propulsion Laboratory; (bottom) Jim Lamb; 73 (top & bottom) NASA/Jet Propulsion Laboratory; 74–75 (top) Jim Lamb; 74–75 (bottom) NASA/Jet Propulsion Laboratory; 76 (top) Ames Research Center/U.S. Geological Survey/ Massachusetts Institute of Technology; (bottom) Jim Lamb; 77 NASA; 78 (left) Jim Lamb; (right) Kitt Peak National Observatory, © 1978 Lowell Observatory & AURA Inc.; 79–81 NASA/Jet Propulsion Laboratory.

Section III: p. 82–83 Goddard Space Flight Center/High Altitude Observatory; 84 Observatoires du Pic du Midi et de Toulouse; 85 © AURA Inc., Kitt Peak National Observatory; 86 Jim Lamb; 87 Joe Goodwin; 88 (top) Observatoires du Pic du Midi et de Toulouse; (bottom) Jim Lamb; 89 Jim Lamb; 90 (top) Jim Lamb; (bottom) "End of the World" by Chesley Bonestell, courtesy of Alfred L. Weisbrich, National Air and Space Museum, Smithsonian Institution, © Time/Life; 91 "Super Nova" by Enrique Zuniga Cordero, National Air and Space Museum, Smith-sonian Institution, photo by Charles H. Phillips; (right) "The White Dwarf" by Ludek Pesek, photo by Charles H. Phillips; 92–93 Jim Brandenburg; 94 (top) Alistair B. Fraser; (bottom) photo by Steven Hodge, © Alistair B. Fraser; 95–96 Alistair B. Fraser; 97 (left) Malcolm Lockwood; (right) Loren McIntyre; 98–99 Alistair B. Fraser; 101 American Science & Engineering/Harvard College Observatory; 102–105 Jim Lamb; 107 P. Mizera, The Aerospace Corp., Space Sciences Laboratory; 108–109 Special Collections Branch, Smithsonian Institution Libraries, photos by Charles H. Phillips; 110 Jim Brandenburg; 111 (top) Kunsthistorisches Museum, Vienna; (bottom) Annie Griffiths; 112 Charles H. Phillips; 113 Tom & Pat Leeson; 114 (top) © 1980, David Muench; (bottom) Bowring Cartographic Research & Design; 115 Glenn Van Nimwegen.

Section IV: p. 116–117 Courtesy of The Reading Public Museum, Pa.; 118–119 Victor Krantz; 119 *Kunstformen der Natur* by Ernst Haeckel, Smithsonian Institution Libraries, photo by Charles H. Phillips; 120–121 Jim Lamb; 122–125 Ross Chapple; 125 (right) *Kunstformen der Natur* by Ernst Haeckel, Smithsonian Institution Libraries, photo by Charles H. Phillips; 126 Glenn Van Nimwegen; 127 Gary Braasch; 128–129 Jim Lamb; (right and far right) from A.J. Hodge, in J.L. Onceley *et al* (eds.), *Biophysical Science— A Study Program,* N.Y., John Wiley & Sons Inc., © 1959 and Gerard J. Tortora *et al* (eds.), *Plant Form and Function,* The Macmillan Company, © 1970; 130 Glenn Van Nimwegen; 131 E.S. Barghoorn; 132 (left) Kjell B. Sandved; 132 (right) Athanasius Kircher, *Magnes sive de Arte Magnetica Opus Tripartitum,* Special Collections Branch, Smithsonian Institution Libraries; 133 (top) Phil Jordan; (bottom) Charlie E. Rogers; 134 (left) Nathaniel B. Ward, *On the Growth of Plants in Closely Glazed Cases,* The John Crerar Library, Chicago, photo by Doug Munson; (right) Bowring Cartographic Research & Design; 135 NASA; 136 (top left) Edward S. Ayensu; (top right) Gary Ladd; (bottom) Kjell B. Sandved; 137 (top & bottom right) Edward S. Ayensu; (bottom left) Kjell B. Sandved; 138–140 Loren McIntyre; 140–141 Edward S. Ayensu; 142

Joe Rychetnik/Photo Researchers, Inc.; 143 (left) Tom Bledsoe/Photo Researchers, Inc.; (right) Ted Spiegel/Black Star; 144 Jim Lamb; 145 *Picturesque Botanical Plates of the New Illustration of the Sexual System of Linnaeus,* 1799, by Robert John Thornton, George Peabody Dept., Enoch Pratt Free Library, Baltimore; 146 Russ Kinne/Photo Researchers, Inc.; 147 Phil Jordan; 148 M. Philip Kahl, Jr./Photo Researchers, Inc.; 149 Glenn Van Nimwegen; 150 Ron Larson; 151 (top) Bob Davis/Woodfin Camp & Associates; (bottom) Kjell B. Sandved; 152 (left) Bill Curtsinger/Photo Researchers, Inc.; (right) Chuck Nicklin/Woodfin Camp, Inc.; 153–155 John Douglass; 156 (top) Glenn Van Nimwegen; 156–157 Michael Friedel/Woodfin Camp & Associates; 157 Kjell B. Sandved.

Section V: p. 158–159 Courtesy of Danish National Museum, photo by Lennart Larsen; 160–161 Courtesy of the Trustees of the British Museum, photo by Michael Holford; 162 Fred J. Maroon; 163 Jonathan Blair/Woodfin Camp, Inc.; 164 (left) Brian Brake/RAPHO, Photo Researchers, Inc.; (right) Fred J. Maroon; 165–166 Fred J. Maroon; 166–167 Robert Lackenbach/Black Star; 168 Ray A. Williamson; 169 *The Bulletin of the Bureau of American Ethnology,* Vol. 135, National Anthropological Archives, Smithsonian Institution, photo by Ray A. Williamson; 170 (left) Kal Muller/Woodfin Camp & Associates; (right) Edward S. Curtis, 1925, National Anthropological Archives, Smithsonian Institution; 171 Jim Lamb, adapted from *The Journal of the Royal Anthropological Institute of Great Britain and Ireland,* Vol. LXI; 172 (left) Kal Muller/Woodfin Camp, Associates; (top) Ray A. Williamson; (bottom) John Running/Black Star; 173 Ray A. Williamson; 174 (left) Schroeder/Eastwood; (right) Glenn Van Nimwegen; 175 California Academy of Science, Owings Collection, photo by Susan Middleton; 176 James A. Sugar/Woodfin Camp, Inc.; 177 (top) Derek J. De Solla Price; (bottom) Collection, American Philosophical Society, Philadelphia, photo by Owen Gingerich; 178–179 Sundial (far right) National Museum of American History, Smithsonian Institution, all others private collection, photo by Charles H. Phillips; 180–181 National Museum of American History, Smithsonian Institution; 180–181 (top) photos by Charles H. Phillips, all others Steve Tuttle; 182 (left) Rosenborg Castle, Copenhagen; (right) Bibliothèque Royale Albert, Ier Brussels; 183 Collection, Caramoor Center for Music and Arts, Katonah, N.Y., photo by Wirtz-Voss Graphics; 184 Museum Folkwang, Essen, Federal Republic of Germany; 185 The Tate Gallery, London; 186 Los Angeles County Museum of Art; 187 © SPADEM, Paris/VEGA, N.Y., courtesy Musees Nationaux, Paris; 188 National Gallery of Art, Washington, D.C., Chester Dale Collection; 189 Courtesy of The Art Institute of Chicago; 190–191 University of Oslo, photo by O. Vaering; 192 National Museum Vincent van Gogh, Amsterdam; 193 The Museum of Modern Art, N.Y.; 194 John Huehnergarth; 195 Photo Giraudon, Paris, © SPADEM, Paris/VEGA, N.Y.; 196–199 John Huehnergarth; 201 © Robert Censoni.

Section VI: p. 202–203 Ross Chapple; 204–205 Martin Rogers/Woodfin Camp, Inc.; 206 Courtesy of the Trustees of the British Museum, photo by Michael Holford; 207 (top) Linda Bartlett/Woodfin Camp, Inc.; 207 (bottom)–208 (top) Tor Eigeland/Black Star; (bottom left) National Gallery of Art, Washington, D.C., Widener Collection; (right) Photo Giraudon, Paris, © SPADEM, Paris/VEGA, N.Y., courtesy Musée Condé; 209 (top) Georg Agricola, *De Re Metallica,* Special Collections Branch, Smithsonian Institution Libraries, photo by Charles H. Phillips; (bottom) M.P.L. Fogden/Bruce Coleman Inc.; 210 The Brooklyn Museum; 211 (left) Thomas Nebbia/Woodfin Camp & Assoc. (right) Tor Eigeland/Black Star; 212 (top) Glenn Van Nimwegen; (bottom) "Sun Maid" and "Sun-Maid Girl Picture Design," registered trademarks of Sun-Maid Growers of California; 213 Aluminum Company of America; 214–225 Pierre Mion; 226–227 Ross Chapple; 228 Pierre Mion; 229 (top) Ira Wexler; (bottom) Solarex Corporation, photo by Fred Leavitt; 230–235 Pierre Mion; 236–237 Art by Jay Mullins for Boeing Aerospace, National Air & Space Museum, Smithsonian Institution, photo by Charles H. Phillips; 237 (right) Bowring Cartographic Research & Design; 238 Art by Pierre Mion, National Air & Space Museum, Smithsonian Institution; 239–240 Art by Barron Storey (detail), NASA, photo by Charles H. Phillips; 241 Art by James Olson for Boeing Aerospace, National Air & Space Museum, Smithsonian Institution, photo by Charles H. Phillips; 242 Art by Barron Storey (detail), NASA, photo by Charles H. Phillips; 243 Courtesy of Peter Glaser/Arthur D. Little, Inc.

Epilogue: p. 244 Walters Art Gallery, Baltimore, photo by Charles H. Phillips; 245 Courtesy of the Trustees of the British Museum, photo by Michael Holford; (right) Pierpont Morgan Library, N.Y.; 246 Benaki Museum, Athens; 247 Schimmel Collection, N.Y.; 248 John Launois/Black Star; (right) Freer Gallery of Art, Smithsonian Institution; 249 Art by Utagawa Kunisada, 1857, Victoria & Albert Museum, London, Crown Copyright, photo by Michael Holford; 250 Linda Bartlett/Woodfin Camp, Inc.; 251 Michael Friedel/Woodfin Camp & Assoc.; 252–253 Tor Eigeland/Black Star; 254–255 Martin Rogers/Woodfin Camp, Inc.; 263 James Tallon.

Front Matter Illustrations

P. 1 Solar arches. 2–3 Center of a sunflower. 4–5 Blue corduroy waves off Hawaii suggest the solar influence on winds, tides, and life in the sea. 6–7 Great Pyramid at Giza, a reminder of the sun-centered culture of ancient Egypt. Title pages: South American dawn silhouettes workers returning home from a night of digging in seaside salt pans. Contents pages: Time-lapse imagery of activity near the sun's surface.

Index

Sun-Earth Comparisons

* 1,300,000 Earths would fit into the sun.

* 333,000 Earths would weigh the same as the sun.

* Hydrogen, the lightest element, and helium make up more than 98% of the sun, while silicon and iron comprise much of the Earth.

* The sun in its present state is at least 4.5 billion years old, and the Earth probably formed before the sun began to shine.

* The sun's diameter of 864,000 miles is 109 times that of Earth.

* Both sun and Earth have magnetic fields, but some of those on the sun are at least 6000 times stronger than Earth's.

* The sun holds the Earth in an orbit roughly 200 million miles in diameter.

* The Earth rotates once in 24 hours, the sun in about 27 days.

* The sun is about 400,000 times brighter than the full moon.

* The sun sheds the light of 3.17×10^{27} candles, but the Earth receives only one part in two billion—or about 1.5 kilowatts per square yard.

* The force of gravity within the solar core is 250 billion times the pull of gravity on the Earth.

* The sun is mostly nothing and is really a burning ball of very light gases. It gets denser as you go toward the center. But it is only as heavy as the air we breathe one third the way into the center, and it is not until halfway to the center that it is as dense as water.

* You could fly through a sunspot in an airplane if you could stand the heat.

* The Earth will be here long after the sun stops shining, but life will not.